6th International Conference on Nanomaterials and Materials Engineering (ICNME 2018)

IOP Conference Series: Materials Science and Engineering
Volume 380

Langkawi, Malaysia
23 – 25 March 2018

ISBN: 978-1-5108-6750-5
ISSN: 1757-8981

Printed from e-media with permission by:

Curran Associates, Inc.
57 Morehouse Lane
Red Hook, NY 12571

Printed by Curran Associates, Inc. (2018)

For permission requests, please contact the Institute of Physics at the address below.

Institute of Physics
Dirac House, Temple Back
Bristol BS1 6BE UK

Phone: 44 1 17 929 7481
Fax: 44 1 17 920 0979

techtracking@iop.org

Additional copies of this publication are available from:

Curran Associates, Inc.
57 Morehouse Lane
Red Hook, NY 12571 USA
Phone: 845-758-0400
Fax: 845-758-2633
Email: curran@proceedings.com
Web: www.proceedings.com

TABLE OF CONTENTS

MATERIAL PHYSICS AND MATERIAL MECHANICS

POLYMERS AND COMPOSITES

METALS AND ALLOYS

IOP Conf. Series: Materials Science and Engineering **380** (2018) 011001 doi:10.1088/1757-899X/380/1/011001

Preface

With great pleasure, we introduce the Proceedings of 2018 6th International Conference on Nanomaterials and Materials Engineering (ICNME 2018) which provides a lively knowledge sharing forum that reports the latest development of research and application pertaining to Nanomaterials and Materials Engineering. ICNME 2018 offers an extensive program of interest to academia, government and industry. Great presentations have been delivered that allow effective knowledge dissemination, as well as enhancing relevant professional practices and skills.

The objective of ICNME 2018 will always be bringing together a large group of researchers, scientists, academics, and engineers in the area of Nanomaterials and Materials Engineering from all over the world to share ideas not only about the technical interests, but also cross culture and history.

This proceedings of ICNME 2018 presents the accepted papers, where the submitted papers are from universities, research institutes and industries. The submitted papers have been gone through a rigorous review and the papers are selected based on quality and relevancy to the conference. The volume tends to present to the readers the recent advancement in the field of Nanomaterials and Nanotechnology; Biomaterials and Biomedical Engineering; Ceramic and Glass Materials; Computational Material Science; Coupling of Multiple Ferroic Order Parameters; Crystal Defect Engineering; Crystallographic Domain Engineering; etc.

We would like to thank all the authors who have contributed to this volume and also to reviewers, speakers, chairpersons, sponsors and all the conference participants for their commitment, hard work and support to ICNME 2018.

ICNME 2018 CONFERENCE CHAIR
March 30, 2018

Conference Committee

INTERNATIONAL ADVISORY COMMITTEES

Prof. Mohamed Henini, University of Nottingham, UK

CONFERENCE CHAIRS

Prof. Jamaluddin Mahmud, Universiti Teknologi MARA, Malaysia

Prof. Brian Yuliarto, Institut Teknologi Bandung, Indonesia

Prof. Dr. Mohd Hamdi Abdul Shukor, University of Malaya, Malaysia

Technical Program Chairs

Prof. Guneik Jang, Chungbuk National University, Korea

Prof. Tjokorda Gde Tirta Nindhia, Udayana University, Indonesia

Prof. Vladimir Tsepelev, Ural Federal University, Russia

Prof. Seong Soo Choi, Sun Moon University, Korea

Technical Committee Members

Prof. Radmila Tomovska, University of the Basque Country, Spain

Prof. Anders Gåård, Karlstad University, Sweden

Prof. Abdul Razak Ismail, Universiti Teknologi Malaysia, Malaysia

Prof. Aidy Ali, National Defence University of Malaysia, Malaysia

Prof. Jonie Tanijaya, Paulus Christian University of Indonesia, Indonesia

Prof. Gunawarman, Andalas University, Indonesia

Prof. Arnold Lubguban, Mindanao State University, Philippines

Prof. Geetesh Goga, K.C. College of Engg. & I.T., India

Prof. S. Balasivanandha Prabu, Anna University, India

Dr. Anita Sagadevan Ethiraj, VIT University, India

Prof. Osman Adiguzel, Firat University, Turkey

Prof. Abbas Teimouri, Payame Noor University, Iran

Assoc. Prof. Mohd Ani, International Islamic University of Malaysia, Malaysia

Dr. Syed Bahari Ramadzan Syed Adnan, University of Malaya, Malaysia

Dr. Azmah Hanim Mohamed Ariff, Universiti Putra Malaysia, Malaysia

Dr. Ping Xiang, City University of Hong Kong, Hong Kong

Dr. Abid Ali Shah, International Islamic University, Pakistan

IOP Conf. Series: Materials Science and Engineering **380** (2018) 011002 doi:10.1088/1757-899X/380/1/011002

IOP Conf. Series: Materials Science and Engineering **380** (2018) 011003 doi:10.1088/1757-899X/380/1/011003

Peer review statement

All papers published in this volume of *IOP Conference Series: Materials Science and Engineering* have been peer reviewed through processes administered by the proceedings Editors. Reviews were conducted by expert referees to the professional and scientific standards expected of a proceedings journal published by IOP Publishing.

IOP Conf. Series: Materials Science and Engineering **380** (2018) 012001 doi:10.1088/1757-899X/380/1/012001

Experimental study of the attenuation properties of metal-viscoelastic bi-layered materials

R F Ghachi and W I Alnahhal

Department of Civil and Architectural Engineering, Qatar University, Doha, Qatar

Abstract. Phononic crystals (PCs) are multi-layered materials with functional elastic wave band gaps where propagation of vibration within these band gaps is restricted. PCs have vibration attenuation properties dependent on their periodic structure and material constituent. In this study, the band gap and vibration attenuation of a one-dimensional PC subjected to longitudinal vibration were evaluated experimentally. Two bi-layered specimens composed of Aluminum and a Silicone Rubber (Elite double 8) were manufactured and tested. The specimens were subjected to vibration from an electrodynamic shaker to obtain pseudo-transfer functions.

1. Introduction

Extreme applications in construction drove to a high demand on engineered materials that have enhanced mechanical proprieties such as high-energy absorbance, vibration control, and acoustic shielding. Mechanical metamaterials (MMs) incorporating viscoelastic constituents hold a great promise to solve extreme loadings problems. The impetus for the incorporation of viscoelastic constituents in MMs mainly stems from their strong energy absorption property, their natural damping and the possession of a low elastic modulus which allows reaching practical low-frequency band-gaps (<20 *k*Hz) that human body is highly affected by. Several researchers introduced simple rheological models (e.g., Kelvin model [1, 2], Zener model [3-5], generalized Maxwell model [6]) in order to simulate material damping pertinent to the bulk material dissipation of MMs. Many studies numerically observed that viscoelasticity not only attenuates wave transmission but also modifies the frequency band-gaps which are substantially displaced and widened [1-6].

Compared to existing and significant numerical research efforts targeting linear viscoelastic MMs, only a limited number of experimental studies [3, 7-11] have validated the aforementioned numerical models. The paucity of experimental results is mainly attributed to the complexities related to material viscosity.

In this paper, we will discuss the experimental results of the electrodynamic shaker test on bi-layered specimens composed of a metallic component: Aluminum and a viscoelastic component: Silicone rubber (Elite double 8).

2. Methodology

The polymers have been mixed and cast in the Structural Laboratory at Qatar University, as illustrated in Figure 1.

IOP Conf. Series: Materials Science and Engineering **380** (2018) 012001 doi:10.1088/1757-899X/380/1/012001

Figure 1. Casting of silicone rubber (Elite double 8) after vigorous mixing.

Cylindrical specimens have been manufactured using cylinders of Aluminum and Elite double 8. Two specimens have been tested. Specimen "A" had a unit cell of 10 mm of Aluminum and 10 mm of Elite double 8, while specimen "B" had a unit cell of 5 mm of Aluminum and10 mm of Elite double 8. The experimental set up for the vibration test is shown in Figure 2. The specimens were placed on a circular base connected to the force transducer and were attached to an accelerometer at the top. Consequently, the force transducer and accelerometer supplied the input and the output signals of the specimen, respectively. These linear perturbation tests were performed by applying chirp sine sweep signal to the specimen up from 1 Hz to 20 kHz. The pseudo-transfer functions were determined from the fast Fourier transform (FFT) ratio of acceleration and force signal as below:

$$FRF(\omega) = \frac{|A_{out}\ (\omega)|}{|F_{in}(\omega)|} \quad (1)$$

where $|A_{out}\ (\omega)|$ and $|F_{in}\ (\omega)|$ represent the frequency spectrum of acceleration and force signals, respectively.

Figure 2. Vibration shaker experiment set-up for pseudo-transfer function test (From top: Accelerometer, Specimen, Force transducer, Shaker).

IOP Conf. Series: Materials Science and Engineering **380** (2018) 012001 doi:10.1088/1757-899X/380/1/012001

3. Results and discussion

The input and output signals for specimen "A" measured by the force transducer and the accelerometer, respectively, are shown in Figure 3.

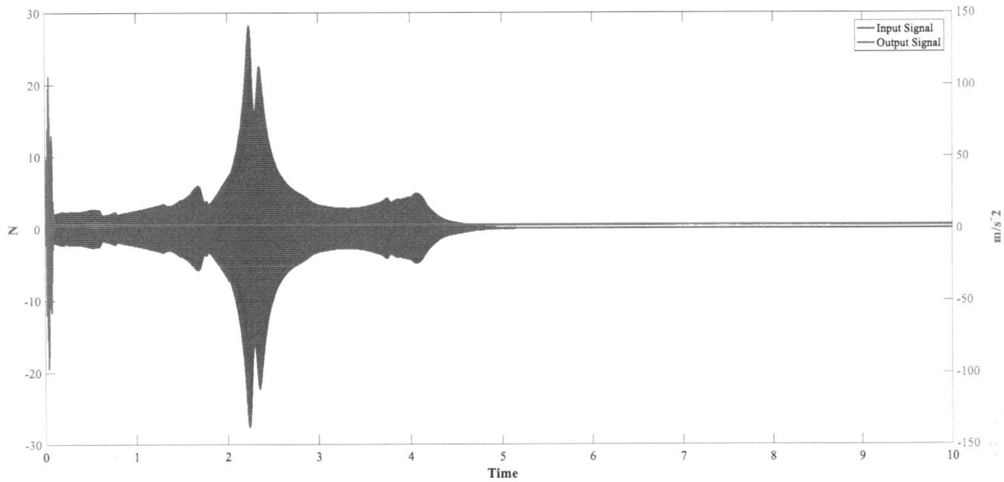

Figure 3. The input signal (blue) and the output signal (green) for specimen "A".

The pseudo-transfer function for specimen "A" is shown in Figures 4. It can be observed that there was a drop in the frequency response at 415 Hz. Then, there was a plateau beyond the frequency of 544 Hz for specimen "A". Observing this FRF of the periodic material, we could see the formation of the bandgap in the frequency range of 0-1500 Hz.

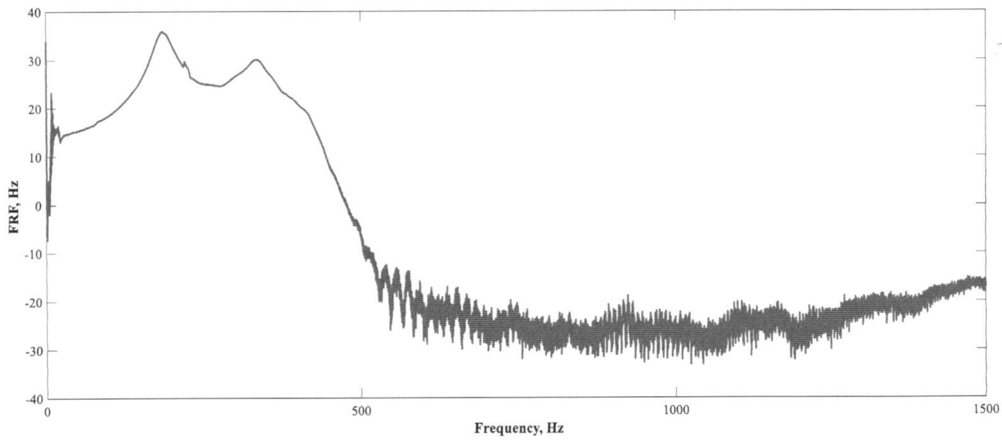

Figure 4. Frequency response function of the specimen "A".

The input and output signals for specimen "B" measured by the force transducer and the accelerometer, respectively, are shown in Figure 5.

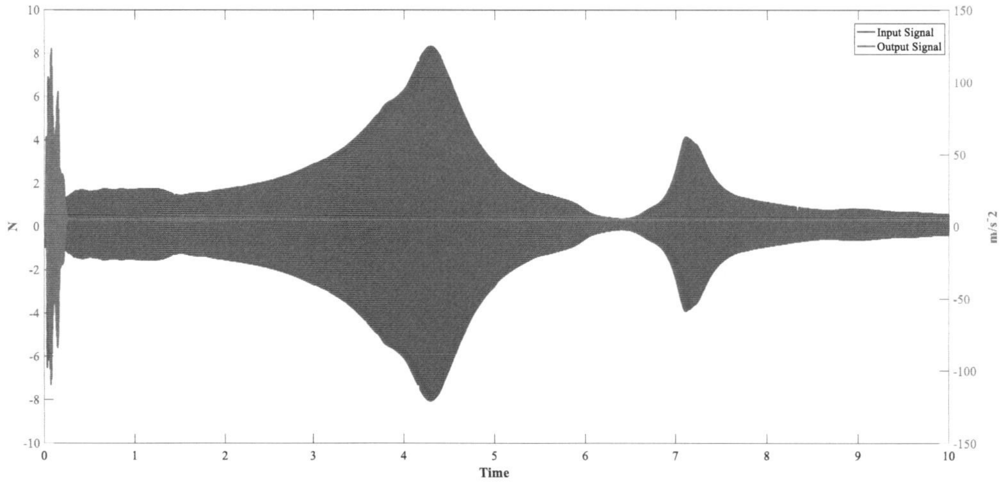

Figure 5. The input signal (blue) and the output signal (green) for specimen "B".

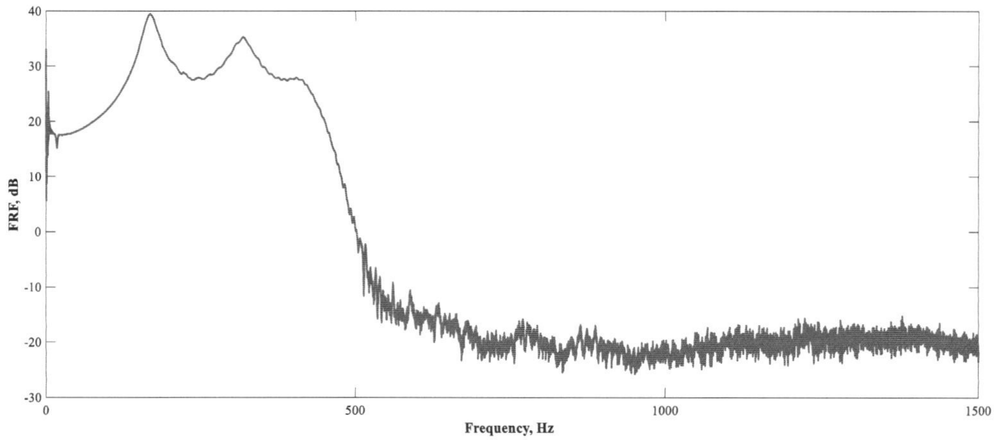

Figure 6. Frequency response function of the specimen "B".

The pseudo-transfer function for specimen "B" is shown in Figures 6. It was observed that there was a drop in the frequency response at 407 Hz. Then, there was a plateau beyond the frequency of 500 Hz for specimen "B". Investigating the FRFs in Figures 4 and 6, one can find the attenuation in the frequency range start earlier with increasing the viscoelastic material ratio.

4. Conclusion

Phononic crystals with a viscoelastic component have been experimentally investigated in this paper. PCs have vibration attenuation properties dependent on their periodic structure and material constituent. In this study, the band gap and vibration attenuation of a one-dimensional PC subjected to longitudinal vibration were evaluated experimentally. Two bi-layered specimens composed of Aluminum and a Silicone Rubber (Elite double 8) were manufactured and tested. The specimens were subjected to vibration from an electrodynamic shaker to obtain pseudo-transfer functions. Experimental test results showed a clear drop in frequency response at 415 Hz and 407 Hz for specimen “A” and “B”, respectively. That was followed by a plateau beyond the frequency of 500 Hz. This is an ongoing research effort, additional tests are required in order to determine the parameters that influence the optimal vibration attenuation.

5. Acknowledgments

The authors show their gratitude to Qatar Foundation for their financial support through an NPRP grant no. NPRP8-1568-2-666.from the Qatar National Research Fund.

6. References

[1] I. E. Psarobas, ''Viscoelastic response of sonic band-gap materials,'' *Phys. Rev.* B, vol. **64**, no. 1, 2001.

[2] Y. Z. Liu, D. L. Yu, H. G. Zhao, J. H. Wen, and X. S. Wen, ''Theoretical study of two-dimensional phononic crystals with viscoelasticity based on fractional derivative models,'' *Journal of Physics D-Applied Physics*, vol. **41**, no. 6, 2008.

[3] B. Merheb, P. A. Deymier, M. Jain, M. Aloshyna-Lesuffleur, S. Mohanty, A. Berker, and R. W. Greger, ''Elastic and viscoelastic effects in rubber/air acoustic band gap structures: A theoretical and experimental study,'' *Journal of Applied Physics*, vol. **104**, no. 6, 2008.

[4] Y. P. Zhao and P. J. Wei, ''The band gap of 1d viscoelastic phononic crystal,'' *Computational Materials Science*, vol. **46**, no. 3, pp. 603–606, 2009.

[5] P. J. Wei and Y. P. Zhao, ''The influence of viscosity on band gaps of 2d phononic crystal,'' *Mechanics of Advanced Materials and Structures*, vol. **17**, no. 6, pp. 383–392, 2010.

[6] B. Merheb, P. A. Deymier, K. Muralidharan, J. Bucay, M. Jain, M. Aloshyna-Lesuffleur, R. W. Greger, S. Mohanty, and A. Berker, ''Viscoelastic effect on acoustic band gaps in polymer-fluid composites,'' *Modelling and Simulation in Materials Science and Engineering*, vol. **17**, no. 7, 2009.

[7] H. G. Zhao, J. H. Wen, D. L. Yu, and X. S. Wen, ''Low-frequency acoustic absorption of localized resonances: Experiment and theory,'' *Journal of Applied Physics*, vol. **107**, no. 2, 2010.

[8] P. Wang, F. Casadei, S. C. Shan, J. C. Weaver, and K. Bertoldi, ''Harnessing buckling to design tunable locally resonant acoustic metamaterials,'' *Physical Review Letters*, vol. **113**, no. 1, 2014.

[9] S. Shan, S. H. Kang, P. Wang, C. Q. Qu, S. Shian, E. R. Chen, and K. Bertoldi, ''Harnessing multiple folding mechanisms in soft periodic structures for tunable control of elastic waves,'' Advanced Functional Materials, DOI:0.1002/adfm.201400665, 2014.

[10] C. Daraio, V. F. Nesterenko, E. B. Herbold, and S. Jin, ''Strongly nonlinear waves in a chain of teflon beads,'' *Phys. Rev.* E, vol. **72**, no. 1, 2005.

[11] H. Meng, J. H. Wen, H. G. Zhao, and X. S. Wen, ''Optimization of locally resonant acoustic metamaterials on underwater sound absorption characteristics,'' *Journal of Sound and Vibration*, vol. **331**, no. 20, pp. 4406–16, 2012.

IOP Conf. Series: Materials Science and Engineering **380** (2018) 012002 doi:10.1088/1757-899X/380/1/012002

Numerical study of the attenuation properties of viscoelastic metamaterials

R F Ghachi and W I Alnahhal

Department of Civil and Architectural Engineering, Qatar University, Doha, Qatar

Abstract. Metamaterials (MMs) are materials engineered to have a property that is not found in nature. In this study, one-dimensional MMs is investigated, particularly that with a functional elastic wave band gaps where propagation of vibration within these band gaps is restricted. The vibration attenuation properties of MMs are dependent on their periodic structure and material constituent. In this paper, the band gap and vibration attenuation of a one-dimensional Phononic crystal (PC) subjected to longitudinal vibration were evaluated numerically. Different geometries of bi-layered specimens composed of Aluminum and a silicone rubber (Elite double 8) were modeled using Finite Element (FE) Software ABAQUS. The FE analysis results were validated using experimental results from the electrodynamic shaker.

1. Introduction

A number of researchers have studied numerically the behavior of phononic crystals (PCs) made of periodic structures designed to control the mechanical waves through Bragg scattering [1-4], locally-resonant materials (LRMs) having the added feature of local resonances [5-8], and cloak materials characterized by transformation-based cloaking/waveguide feature [9-13]. Potential applications of MMs can range from vibration control [14], noise reduction [15], to waveguide endeavors [16].

A periodic mass-spring system was created by Wen et al. [17] to determine the band gap property of the crystal. The set of round copper and aluminum plates were connected by spring through a guiding pole system. The crystal was excited along the vertical direction by a vibrometer (VE 5000) derived by signal analyzer (B&K 3650). For the excitation signal in the frequency range of 0 to 2 kHz, the response was obtained by an accelerometer (B&K 4507B) at the base and top of the crystal. Comparison of the two signals presents the attenuation characteristics of the crystal. The crystal was analyzed for the loaded and unloaded condition. In both cases, attenuation of the wave was found at frequencies of 60-125 Hz.

Another one-dimensional PC was experimentally investigated by Yan-Lin et al. [18]. Only the longitudinal wave was applied to the specimen made of copper and silicone rubber material through a vibration exciter. The diameter and length of each plate were 25 mm and 20 mm, respectively, which gives a total 120 mm long specimen. A force sensor was used for measuring the applied force of the vibration and an accelerometer was used for determining the dynamic response. The experiment was performed using PUMA vibration control system. The result was obtained for a low-frequency range of 0 to 2 kHz and a definite bandgap frequency range was not observed in that range. The reason was attributed to the viscoelastic damping property of the rubber material.

In this paper, we will discuss the numerical results of the electrodynamic shaker test on bi-layered specimens composed of a metallic component: Aluminum and a viscoelastic component: Silicone rubber (Elite double 8).

2. Methodology

An asymmetrical model of Aluminum and Elite double 8 was modeled to get its transfer function using a steady state analysis using FE Software ABAQUS, as shown in Figure 1.

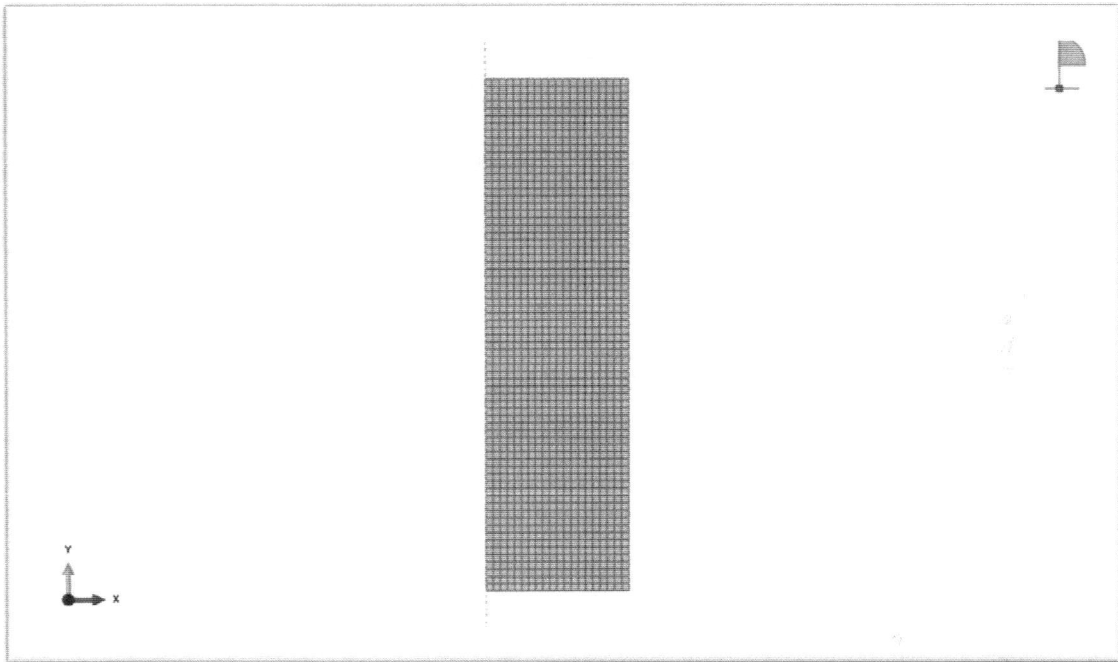

Figure 1. Asymmetrical model of Aluminum and Elite double 8 bi-layered material

The numerical model simulates two cylindrical specimens that have been tested using an electrodynamic shaker. Specimen "A" had a unit cell of 10 mm of Aluminum and 10 mm of Elite double 8, while specimen "B" had a unit cell of 5 mm of Aluminum and10 mm of Elite double 8. The experimental set-up for the vibration test is shown in Figures 2. The specimens were placed on a circular base connected to the force transducer and were attached to an accelerometer at the top. Consequently, the force transducer and accelerometer supplied input and output signal of the specimen, respectively.

Figure 2. Vibration Shaker experiment set up for pseudo-transfer function test (From top: Accelerometer, Specimen, Force transducer, Shaker).

The FE models were used to predict the output signal for both specimens "A" and "B". Then, the transfer functions for both specimens were computed and compared to the experimental result in Figures 3 and 4.

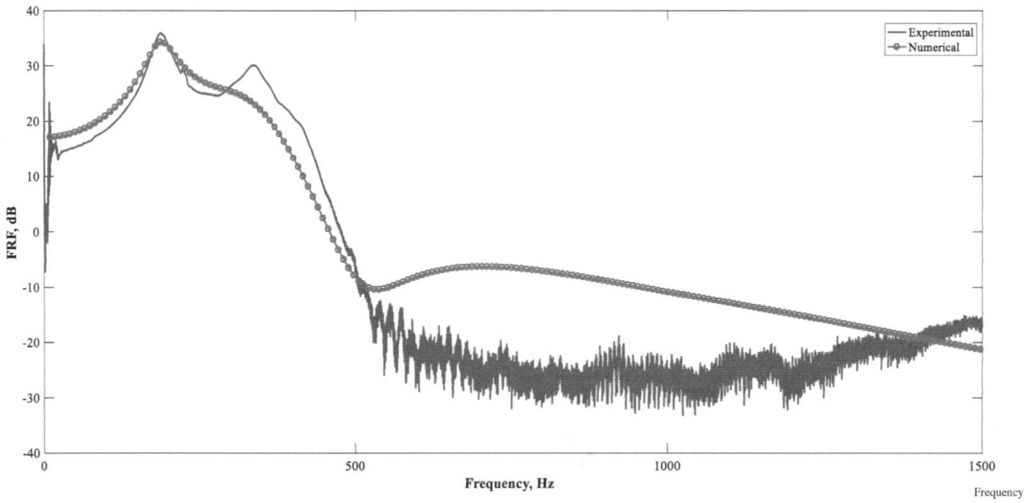

Figure 3. The pseudo transfer function plot for specimen "A" (a) the experimental FRF [Solid line] (b) The finite element model FRF [Dotted line].

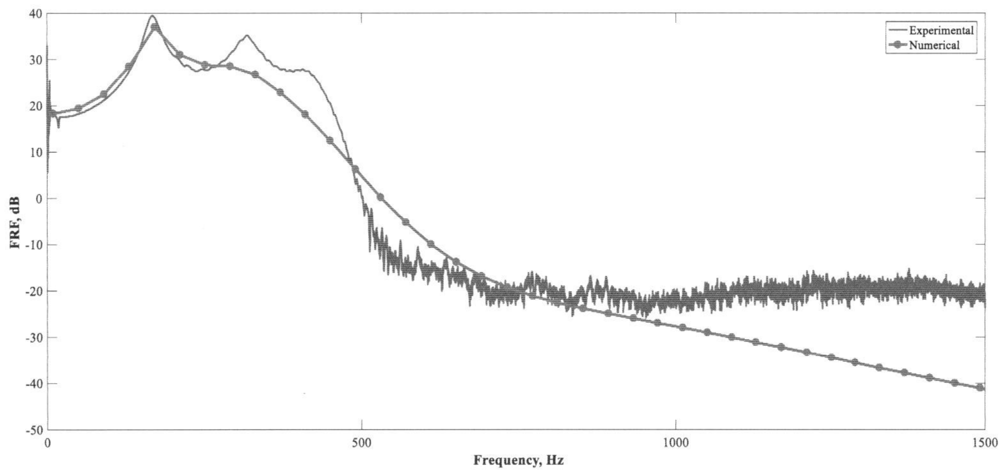

Figure 4. The pseudo transfer function plot for specimen "B" (a) the experimental FRF [Solid line] (b) The finite element method FRF [Dotted line].

The plots in Figures 3 and 4 represent a comparison between the experimentally measured and the numerically calculated pseudo transfer function for specimens "A" and "B", respectively. Although the threshold frequency at which the attenuation begins visibly by the clear drop in the frequency response is close between the experimental and numerical results. The attenuation behavior of the MM specimen was not consistent with the FEM version of FRF.

3. Conclusion

One-dimensional MM composed of alternate aluminum and rubber layers is considered in this study. The primary goal of using viscoelastic MM is to acquire band-gap behavior below the frequency range of common human activity (<20 kHz). An asymmetrical model was simulated using FEM to obtain the predicted FRF of a bi-layered material designed for longitudinal vibration attenuation. Noticeably, the attenuation result of the MM specimen was not consistent with the elastic version of FRF. This result suggests that the nonlinear viscoelastic behavior of the polymeric materials should be considered for an accurate estimation of attenuation property of MM. Proper steps will be taken in this ongoing research to report on an adequate damping model.

4. Acknowledgments

The authors show their gratitude to Qatar Foundation for their financial support through an NPRP grant no. NPRP8-1568-2-666.from the Qatar National Research Fund.

5. References

[1] M. Sigalas and E. N. Economou, ''Band-structure of elastic-waves in 2-dimensional systems,'' *Solid State Communications*, vol. 86, no. 3, pp. 141--143, 1993.
[2] M. S. Kushwaha, P. Halevi, L. Dobrzynski, and B. Djafarirouhani, ''Acoustic band-structure of periodic elastic composites,'' *Physical Review Letters*, vol. **71**, no. 13, pp. 2022--25, 1993.
[3] M. Ruzzene and F. Scarpa, ''Directional and band-gap behavior of periodic auxetic lattices,'' *Physica Status Solidi B-Basic Solid State Physics*, vol. **242**, no. 3, pp. 665--680, 2005.
[4] M. Maldovan and E. L. Thomas, ''Simultaneous complete elastic and electromagnetic band gaps in periodic structures,'' *Applied Physics B-Lasers and Optics*, vol. **83**, no. 4, pp. 595--600, 2006.

[5] J. R. Willis, ''A polarization approach to the scattering of elastic-waves .2. multiple-scattering from inclusions,'' *Journal of the Mechanics and Physics of Solids*, vol. **28**, no. 5-6, pp. 307--327, 1980.
[6] J. R. Willis, ''Variational and related methods for the overall properties of composites,'' *Advances in Applied Mechanics*, vol. **21**, pp. 1--78, 1981.
[7] J. R. Willis, ''Variational-principles for dynamic problems for inhomogeneous elastic media,'' *Wave Motion*, vol. **3**, no. 1, pp. 1--11, 1981.
[8] G. W. Milton and J. R. Willis, ''On modifications of Newton's second law and linear continuum elastodynamics,'' Proceedings of the Royal Society a-Mathematical Physical and Engineering Sciences, vol. **463**, no. 2079, pp. 855--880, 2007.
[9] G. W. Milton, M. Briane, and J. R. Willis, ''On cloaking for elasticity and physical equations with a transformation invariant form,'' *New Journal of Physics*, vol. **8**, 2006.
[10] A. N. Norris, ''Acoustic cloaking theory,'' Proc. of the Royal Society a-Mathematical Physical and Engineering Sciences, vol. 464, no. 2097, pp. 2411--24, 2008.
[11] H. Y. Chen and C. T. Chan, ''Acoustic cloaking and transformation acoustics,'' *Journal of Physics D-Applied Physics*, vol. **43**, no. 11, 2010.
[12] S. Zhang, C. G. Xia, and N. Fang, ''Broadband acoustic cloak for ultrasound waves,'' *Physical Review Letters*, vol. **106**, no. 2, 2011.
[13] T. Buckmann, M. Thiel, M. Kadic, R. Schittny, and M. Wegener, ''An elasto-mechanical unfeelability cloak made of pentamode metamaterials,'' *Nature Communications*, vol. **5**, 2014.
[14] F. Casadei, B. S. Beck, K. A. Cunefare, and M. Ruzzene, ''Vibration control of plates through hybrid configurations of periodic piezoelectric shunts,'' *Journal of Intelligent Material Systems and Structures*, vol. **23**, no. 10, pp. 1169--1177, 2012.
[15] F. Casadei, L. Dozio, M. Ruzzene, and K. A. Cunefare, ''Periodic shunted arrays for the control of noise radiation in an enclosure,'' *Journal of Sound and Vibration*, vol. **329**, no. 18, pp. 3632--46, 2010.
[16] J. O. Vasseur, A. C. Hladky-Hennion, B. Djafari-Rouhani, F. Duval, B. Dubus, Y. Pennec, and P. A. Deymier, ''Waveguiding in two-dimensional piezoelectric phononic crystal plates,'' *Journal of Applied Physics*, vol. 101, no. 11, 2007.
[17] J. H. Wen, G. Wang, D. L. Yu, H. G. Zhao and Y. Z. Liu, "Study on the vibration band gap and vibration attenuation property of phononic crystals," *Science in China Series E: Technological Sciences*, vol. 51, no. 1, pp. 85-99, 2008.
[18] W. Yan-Lin, C. Ming-Wen and W. Zi-Dong, "Study on band gap structure of one dimensional phononic crystal," *Advanced Materials Research*, Vols. 152-153, pp. 1696- 99, 2010.

IOP Conf. Series: Materials Science and Engineering **380** (2018) 012003 doi:10.1088/1757-899X/380/1/012003

Capacitance-voltage characteristics of LB thin films incorporating cds nanoparticles

S Malik[1], **A K Ray**[2] **and F H Naning**[3]

[1] Department of Physics, Faculty of Science and Mathematics, Universiti Pendidikan Sultan Idris, 35900 Tanjong Malim, Perak DR, Malaysia
[2] Institute of Materials and Manufacturing, Brunel University, London, Uxbridge, UB8 3PH, UK
[3] Faculty of Agriculture and Food Sciences, Universiti Putra Malaysia, Bintulu Campus, 97008 Bintulu, Sarawak, Malaysia

Abstract. Two types of Langmuir-Blodgett (LB) thin films based on hybrid organic-inorganic materials sandwiched between metal and semiconductor were fabricated. The novel films were 40 layers Y-type LB films of Cd-salt stearic acid ($CdSt_2$). The second type of films was formed after the treatment of $CdSt_2$ films with H_2S gas over a period of 12 hours at room temperature to grow CdS nanoparticles within the stearic acid matrix. The capacitance-voltage (C-V) measurement of $CdSt_2$ LB films exhibit a significant dependence on the measurement frequency in the accumulation region due to high DC leakage currents. By embedding CdS nanoparticles into the stearic acid matrix, less frequency dependent C-V curves were obtained. The problem in determining the true insulator capacitance due to frequency dispersion was overcome by using the Yang's model. The corresponding dielectric constant of LB films of $CdSt_2$ was found to be 2.3 and increased to 5.1 when embedded with CdS nanoparticles.

1. Introduction

Silicon wafer is widely dominating the microelectronics industry due to its stability and low manufacturing cost, making it technologically an important substrate. By depositing organic thin films directly on top of the silicon surface, the formed electron devices can be tuned through precise control of the insulating layers. Central to the application of organic films in electron devices is the electrical properties of the monolayers. By embedding semiconductor nanoparticles such as cadmium sulphide (CdS) into the organic films matrix, a device with novel electrical and optical properties is created [1]. The Metal-Insulator-Semiconductor (MIS) structure is an important part of many semiconductor devices [2-3]. The presence of a dielectric material between two conductors gives the device the properties of a capacitor. The capacitance-voltage measurement is a powerful way to investigate the quality of the dielectric and the quality of the insulator-semiconductor interface [4].

The LB-technique is found to be one of the most promising methods for preparing organic thin films [5]. This method is usually used to fabricate highly ordered ultra thin films of various organic materials based on amphiphilic molecules. It has some advantages such as precise control of the monolayer thickness, homogeneous deposition of the monolayer over large areas and the possibility to make multilayer structures with varying layer composition [6]. An additional advantage of the LB technique is that monolayer can be deposited on almost any kind of solid substrate. Nowadays molecular coating on solid surfaces is important in physics, chemistry, material science, and electronics with potential application in optoelectronic devices. On the other hand, C–V analysis is considered as one of the most important tools for characterizing MIS [7] and Metal-Oxide-Semiconductor (MOS) [8] devices.

2. Materials and methodology

The thin films investigated in this study were insulating LB films of stearic acid transferred from the subphase containing metallic cadmium (Cd^{2+}) cations. Composite LB films were deposited onto n-type silicon substrate and MIS structures were fabricated containing these stearic acid (SA) films and will be referred as "untreated" samples. Some LB films of composite stearic acid were then treated with hydrogen sulphide gas in order to introduce cadmium sulphide (CdS) nanoparticles into the untreated matrix. These MIS structure devices will be named as "treated" samples. The fabrication of MIS structures involved careful preparation of silicon (Si) substrates, and deposition of thin organic film and electrical contacts.

2.1. Thin film deposition

LB films were deposited onto a silicon wafer substrate by moving the substrate downwards and upwards through the sub phase surface by vertical lift. The silicon wafer was used as a substrate due to their extremely smooth surface [9]. Substrates can be divided into two categories; hydrophilic and hydrophobic. Deposition onto a hydrophilic substrate (e.g. Al coated glass slide) only begins on the upward motion of the substrate whilst transferring a monolayer onto a hydrophobic surface (e.g. silicon wafer) begins on the downward motion. Figure 1 shows the upward and downward motion of the hydrophilic and hydrophobic substrates respectively.

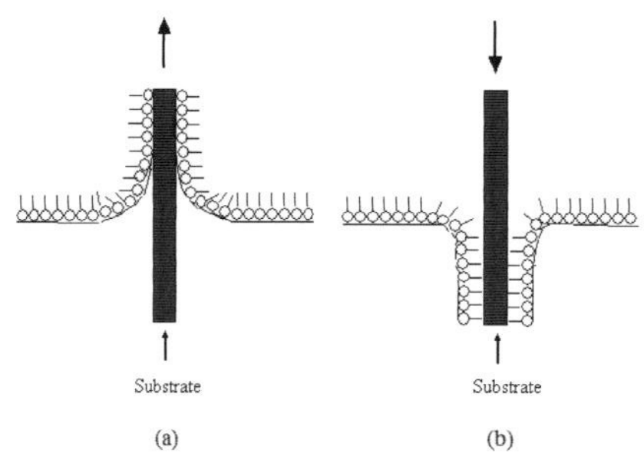

Figure 1. Deposition of first monolayer onto; (a) hydrophilic substrate (upward); (b) hydrophobic substrate (downward)

2.2. Device fabrication

The fabrication of devices used in this study was performed in a clean room (class 100) environment following carefully drawn up procedures including health and safety guidelines. The methods for device fabrication involved etching and cleaning of silicon substrates, preparation of the LB trough, LB film deposition, cadmium sulphide (CdS) nanoparticles formation, and metal contact deposition.
Substrates in this investigation were n-type silicon wafers (100). The thickness and resistivity (ρ) were approximately 300 μm and 4.5 Ω.cm, respectively. As part of the device requirements, silicon dioxide (SiO_2) layers with thickness in the range of 0.8 μm to 1.5 μm were thermally grown on top of the silicon wafers. The wafers were then cut to the dimension of a glass slide which was approximately 76 mm x 26 mm. A part of the SiO_2, approximately 15 mm, was then etched in a mix of hydrofluoric (HF) acid and Millipore water (18 MΩ.cm) with a moderate concentration ratio of 1:5, resulting in 15 mm width of the bare silicon substrate. In order to measure the capacitance characteristics of the film, metal contacts of Aluminium were deposited on both sides of the device. Figure 2 shows the schematic diagram of the fabricated device where A is the affective device area.

IOP Conf. Series: Materials Science and Engineering **380** (2018) 012003 doi:10.1088/1757-899X/380/1/012003

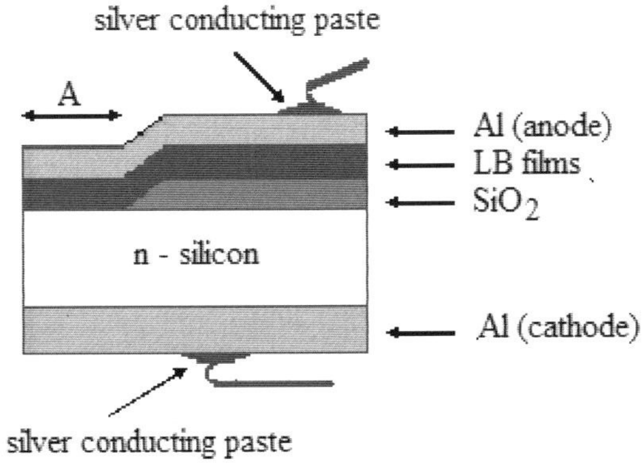

Figure 2. A schematic diagram of the fabricated device.

2.3. Capacitance measurements

Small-signal C-V-f measurements were carried out using Hewlett Packard 4284A Precision LCR meter to measure the capacitance as a function of either frequency (C-f) or DC voltage (C-V). Prior to measurement the LCR meter was calibrated for conductor cables used in connecting the samples by open and short-circuit test. For C-f, measurements were made by applying a small AC signal of $V = 20$ mVrms with frequency varied in a semi-logarithmic preset steps from 20 Hz to 1 MHz at zero DC voltage. For C-V, the same AC signal with a modulating frequency set of several chosen values was superimposed on a DC voltage and swept between ± 3V. The DC bias sweep was $0.01 Vs^{-1}$ with a time delay of 5s and 10s for the DC and AC measurements respectively.

3. Results and discussion

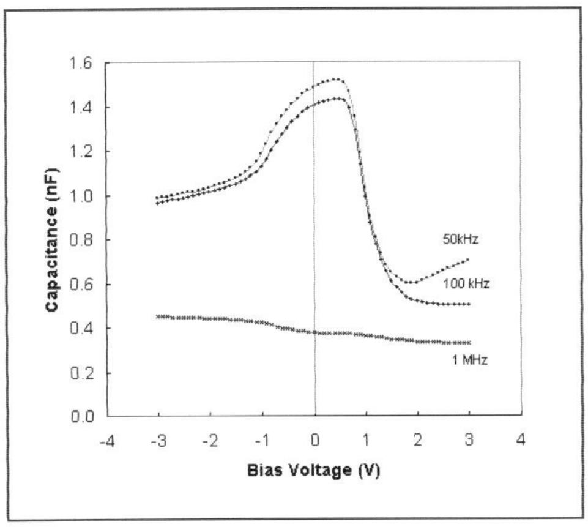

Figure 3. Frequency-dependent capacitance–voltage (*C–V*) characteristics of "untreated" devices at three different frequencies.

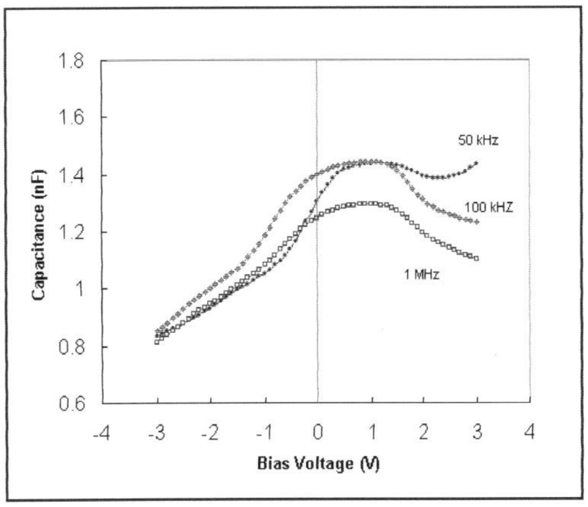

Figure 4. Frequency-dependent capacitance–voltage (*C–V*) characteristics of "treated" devices at three different frequencies.

Figure 3 depicts the C-V characteristics of the untreated device (SA) for three different modulating signal frequencies. As can be seen from the graph, the observed behaviours were different from those obtained from ideal MIS devices. The capacitance of the untreated device was found to be highly dependent on the AC modulating frequency. For AC signal frequency up to 100 kHz, a typical C-V behaviour of MIS structure showing three distinct regions: inversion, depletion and accumulation were observed when the biasing voltage was swept from -3V to up to approximately 0.8V. At high frequency (e.g. 1 MHz) the current carriers are not able to follow the AC signal, resulting in a capacitance value, almost independent of the applied voltage. Similar results were reported for organic/inorganic hetero-diodes [10].

A different behaviour was observed for a treated device. Figure 4 depicts the variation of capacitance of Al/SA+CdS/Si/Al devices as a function of applied voltage for three modulating frequencies, 50 kHz, 100 kHz and 1 MHz. The capacitance showed smaller dependence on the AC signal frequency than untreated samples. An improvement was also achieved in the C-V behaviour. The C-V characteristics were, however, not ideal as expected for MIS capacitor of SiO_x [11] but depletion and accumulation regions were found to be well-defined for all modulating frequencies.

The presence of an inversion layer was not obvious and it was believed to occur at higher reverse voltage. A further decrease of the capacitance with increasing reverse bias voltage indicated that the space-charge layer was driven far into the inversion-layer regime without the production of any inversion-layer charge.

A method presented by Yang and Hu for MOS capacitor with thin SiO_2 dielectric and large DC leakage current [12] was employed in order to determine the true insulator capacitance in our organic LB films. This modelling technique was found to be suitable for the frequency dispersion of C-V measurements in the accumulation region of MOS devices. Figure 5 shows the simulated capacitance as a function of bias voltage. As it can be seen from the graph, both sets of frequencies chosen for calculating the true insulator capacitance were consistent with each other and frequency-independent. Also the treated devices have a larger value of capacitances and better C-V characteristics as compared to the untreated one. Due to flattened capacitance at a frequency of 1 MHz, the simulated capacitances for untreated device did not clearly show the MIS behaviour as compared to the treated one. Nevertheless, we can still observe the flat accumulation region, which is in the region of our interest.

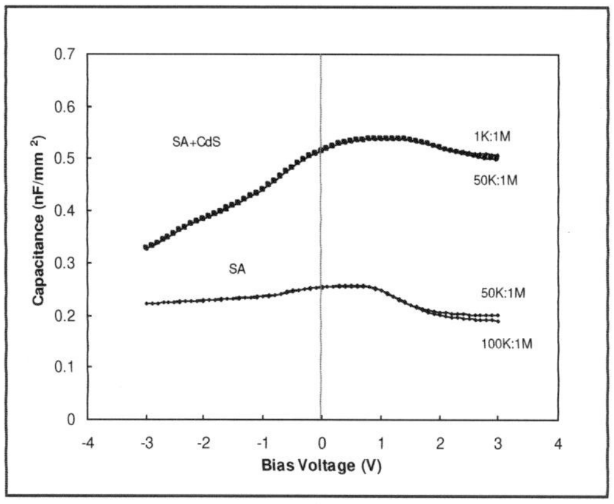

Figure 5. Simulated C-V characteristics obtained by using Yang's technique.

The flat accumulation capacitances taken from the graph in Figure 5 were 1.35 nF and 0.9 nF for treated and untreated devices respectively. The true insulator capacitance was determined when the device was biased in the accumulation region. Taking into consideration the parallel SiO_2 capacitance, the SiO_2 capacitance was estimated to be 0.22 nF for treated and 0.17 nF for untreated device. By subtracting the SiO_2 capacitance, the true capacitance value of LB films was found to be 1.13 nF for treated and 0.73 nF for untreated devices. The corresponding effective dielectric constant (ε_r) is estimated to be 2.3 and 5.1 for the SA film and LB films containing nanoparticles respectively.

4. Conclusion

A study on the capacitance-voltage-frequency behaviour of treated and untreated LB films in MIS structure has been performed in order to determine their dielectric permittivity (ε_{SA+CdS} and ε_{SA}) by measuring their capacitance. The presence of an inversion layer was not obvious and it was believed to occur at higher reverse voltage. A further decrease of the capacitance with increasing reverse bias voltage indicated that the space-charge layer was driven far into the inversion-layer regime without the production of any inversion-layer charge. Due to the anomalous frequency dispersion in the accumulation region, it was not possible to determine the dielectric constant of the LB films. However, by using a two-frequency technique (Yang's model) a true LB film capacitance was extracted from the simulated C-V curve hence, the dielectric constant was obtained. It was found that, the dielectric constant for untreated device has doubled after incorporating cds nanoparticles into the stearic acid matrix. Metal-Insulator-Semiconductor (MIS) or recently known as Metal-Organic-Semiconductor (MOS) materials have established great interest among researchers. The low cost solution-processed deposition technology on large-area substrates at room temperature might be exploited for applications such as thin-film-transistors, solar cells, sensors, and light-emitting diodes.

5. Acknowledgment

S Malik would like to thank Universiti Pendidikan Sultan Idris, Malaysia for financial contributions for this project.

6. References

[1] Sagadevan S 2013 *Nanoscience and Nanotechnology* 3(3) 62-74.
[2] Mark T G and Zheng H L 2013 *NPG Asia Materials* 5 e55.
[3] Tao Z and Meng N C 2015 *Nano Energy* 12 347–373.
[4] Cai L, Zhang S, Miao J, Wei Q and Wang C 2015 *Nanoscale Research Letters* 10(1) 291.
[5] Zhavnerko G and Marletta G 2010*Materials Science and Engineering: B* 169 43-48.
[6] Rohit M, Kriti G, Dhruv T, Vaibhav V, John R M, Neeru S and Annapoorni S 2014 *Surface and Coatings Technology* 258 509-514.
[7] Gerke S, Micard G, Job R, Hahn G. and Terheiden B 2016 *Phys. Status Solidi C* 13 724–728.
[8] Jie B B and Sah C T *Journal of Semiconductors* 2011 32(12) 121001.
[9] Young H, Liao H and Huang H 2006 *Int J Adv Manuf Technol* 29 372-378.
[10] Landi G, Fahrner W R, Concilio S, Sessa L and Neitzert H C 2014 *IEEE Journal of Electron Devices Society* 2(6) 179-181.
[11] Sze S M 2007 *Physics of Semiconductor Devices* 3rd ed. New Jersey: Wiley-Interscience.
[12] Yang K J and Chenming H 1999 *IEEE Transaction on Electron Devices* **46(7)** 1500-1501.

Nickel zinc ferrite thick film as substrate overlay for improved performance of microstrip patch antenna

I H Hasan[1], M N Hamidon[1,2], A Ismail[2], I Ismail[1], M A M Kusaimi[1], S Azhari[1], N A M Azhari[1], and R Osman[1]

[1]Institute of Advanced Technology, University Putra Malaysia, 43400 UPM Serdang, Selangor, Malaysia
[2]Faculty of Engineering, University Putra Malaysia, 43400 UPM Serdang, Selangor, Malaysia

Email: i_helina@upm.edu.my

Abstract. In this paper, nickel zinc ferrite (NZF) nanopowder was mixed with organic vehicle which consists of linseed oil, m-xylene and α-terpineol. Then the mixture was sonicated for 1 hour at 40°C in order to obtain homogenous paste, followed by printing it onto FR4 substrate using the screen printing technique to form the NZF thick film layer before dried at 100°C and later fired at 200°C. A basic square shape patch antenna by using silver paste was printed onto the NZF thick film layer and was compared with another patch antenna which was been printed without the NZF layer. The results showed that the antenna with NZF thick film layer has return loss of -10.97dB, resonant frequency 6.42GHz, bandwidth 3.9 and Q factor of 1.646, which is better compared to the antenna without the layer by 32.81%, 3.22%, 86.60% and 44.69% respectively.

1. Introduction

Thick film technology is the most commonly used technology for producing electronic devices due to its inexpensive production and simple method by using thick film paste. Thick film circuits and devices are fabricated by screen printing active materials in a form of a viscous film paste on ceramic or polymer substrates. Thick film paste plays an important role in thick film technology, since it relies heavily on the characteristics of the paste. One of the ingredients in the paste is the active element which will determine the properties and characteristics of the paste [1]. In recent years, thick film technology has also been adapted to fabricate microstrip patch antennas, in addition to the conventionally used patch antenna fabrication using printed circuit board (PCB) etching technique [2]. Ferrimagnetic materials or ferrites are one type of magnetic materials, having a certain degree of susceptibility to magnetization. Ferrites are divided to two categories, which are hard and soft ferrites. Hard ferrites have high coercivity and high remanence after being magnetized, which makes them perfect as permanent magnets. Soft ferrites on the other hand, have low coercivity that makes the magnetic direction can be easily reversed to normal state with low loss energy. These types of ferrites also have high electrical resistivity and low magnetic losses at higher frequency range, which makes them suitable for microwave applications [3]. Ferrites are also being studied as substrates for microstrip patch antennas (MPA), which can be tuned when external magnetic field is applied [4]. Ferrite bulks have been considered as substrates to improve MPA performance, however it is difficult to control the permittivity and permeability of the magnetic materials, and the fabricated ferrite

IOP Conf. Series: Materials Science and Engineering **380** (2018) 012004 doi:10.1088/1757-899X/380/1/012004

substrates tend to be brittle. Therefore, this work focuses on the use of NZF as thick film layer in MPA fabrication to improve the performance of MPA in terms of bandwidth and return loss.

2. Experimental method

In the paste preparation process, the paste was prepared by mixing 30 wt% of NZF nanopowder (<100nm particle size, 99.9% trace metals basis) with 70 wt% of organic vehicle which consists of linseed stand oil, m-xylene and α-terpineol, and mixed thoroughly using sonicator for 1 hour with 10 minutes interval after 30 minutes to obtain homogenous paste for thick film printing [5]. The prepared thick film paste was then printed onto flame-retardant 4 (FR4) substrate using screen printing method for characterization purpose. The thick film was printed using a silk screen frame with a 20 x 20 mm square design using a screen printing machine. Next, the prepared sample was dried first at temperature 100°C for ten minutes, and later fired at temperature 200°C for 30 minutes in a box furnace (HT4–1600–SIC, Malaysia). The thick film was then characterized using Field Emission Scanning Electron Microscope (FESEM) (Hitachi S-3400N). Permittivity, dielectric loss tangent and permeability of the samples were measured using Impedance Analyzer (Agilent HP4291B) for frequency ranging from 10MHz until 1GHz.

After thick film characterization was completed, fabrication of microstrip patch antenna was done by first printing NZF thick film onto FR4 substrate. After the thick film was dried and fired to ensure adhesion, a 10 x 10 mm square of conductive patch and microstrip transmission line was then screen printed onto the thick film layer using silver paste to form a radiating patch for the antenna, and afterwards dried and fired at 150°C for 30 minutes. The return loss and resonant frequency for all prepared samples with different substrates and parameters were measured using Vector Network Analyzer (Keysight, PNA 5227A). The measurement frequency range is set between 2GHz up until 8GHz. Another patch antenna was also fabricated without NZF thick film as substrate overlay for comparison. Figure 1 shows the structure of the designed antenna and the actual fabricated MPA.

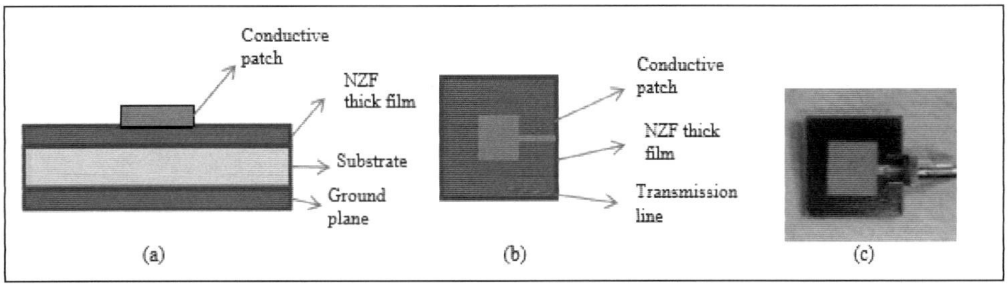

Figure 1. Fabrication of MPA: (a) side view, (b) top view, (c) actual fabricated MPA (top view only).

Figure 2. FESEM images of NZF thick film: (a) before firing, (b) after firing at 200°C.

3. Results and discussion

Figure 2 shows the FESEM images of the NZF thick film before and after firing at 200°C. Before the thick film is fired at a certain temperature, the upper layer of the thick film was covered by the organic vehicle which is in liquid form. When the film is subjected to a certain fixed temperature, the organic vehicle dried off and evaporated while at the same time holding the nanoparticles and bound them to the substrate. Linseed stand oil which is also known as drying oil polymerizes when exposed to a certain temperature. This makes the oil suitable as a vehicle as well as a binder in thick film paste. After firing process is completed, it can be observed from Figure 2(b) that the particles are visible as compared to the thick film before firing. The particles however seemed to agglomerate due to particle size of NZF nanopowders, making the particles attached to each other to minimize energy loss.

Permittivity and permeability of NZF thick film for frequency ranging from 10 MHz to 1 GHz are shown in Figure 3. The permittivity is 2.2, whereas the permeability of the material is 1.12 Based on these results, NZF has low permeability but high permittivity, which makes it a good dielectric material for antenna fabrication.

Finally, Figure 4 shows the return loss or S_{11} parameter and resonant frequency of the fabricated patch antenna compared with MPA without NZF thick film layer. It can be clearly seen that by adding on NZF thick film layer onto the substrate, the return loss and -3dB bandwidth improved, while the resonant frequency (f_r) shifted by only 0.2GHz, which is only 3.22% increment. The return loss improved by 32.81% which is from -8.26dB without NZF thick film and -10.97 with the NZF thick film. The bandwidth (BW) has also improved significantly by 86.60%, which is from 2.09 increased to 3.90. As a result, the Q factor of the antenna which can be derived from equation has decreased from 2.976 to 1.646 due to wider BW.

$$Q = \frac{f_r}{BW} \tag{1}$$

When a magnetic material in included in the volume between a radiating patch and the ground plane, the current amplitude induced becomes smaller due to permeability of the material. When this happens, automatically the energy stored in the volume is decreased, therefore decreasing the quality factor of the antenna. The decreased Q factor will then increase the bandwidth of antenna [6], thus enhance the performance of MPA.

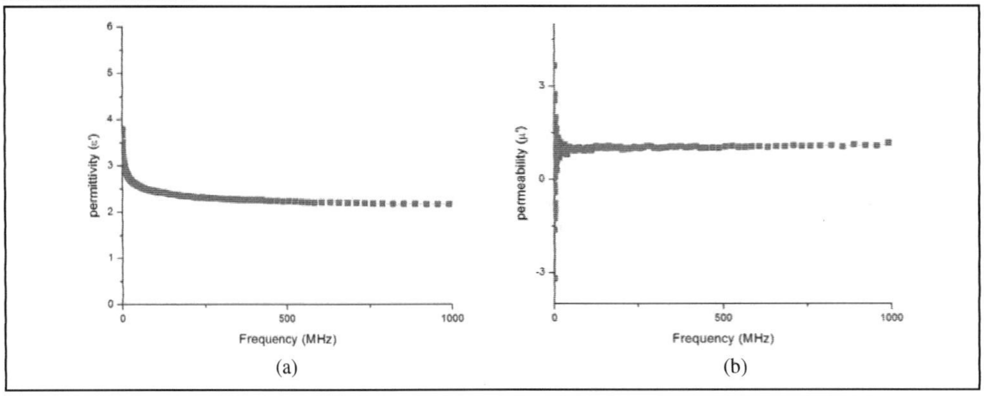

Figure 3. (a) Permittivity and (b) permeability of NZF thick film.

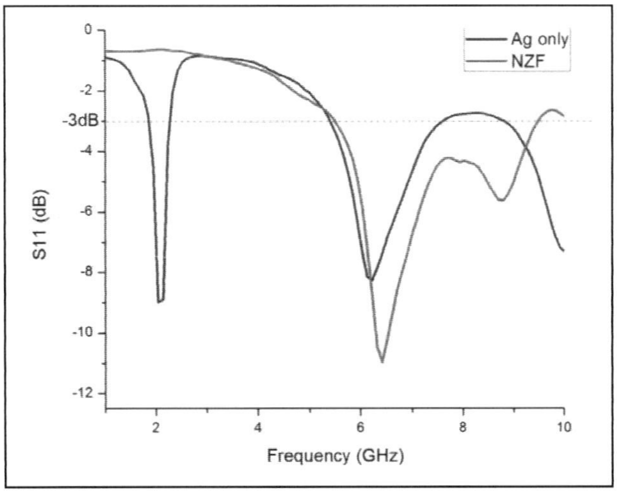

Figure 4. Results of MPA with silver (Ag) patch only vs. MPA with Ag and NZF thick film as substrate overlay.

4. Conclusion

This paper studied the inclusion of NZF thick film as substrate overlay to MPA system to improve the bandwidth and return loss of the MPA. The results showed that the bandwidth improved significantly, therefore decreasing the q factor of the MPA. The work proved that by adding ferrite based thick film layer to MPA during fabrication, the performance of the MPA can be greatly enhanced, specifically the bandwidth of the antenna which has always been the issue when using MPA.

5. Acknowledgments

The authors wish to acknowledge financial support from Ministry of Science, Technology and Innovation (MOSTI) under Sciencefund grant (03-01-04-SF1860), Universiti Putra Malaysia under Putra grant (GP-I/2014/9439400) and Ministry of Higher Education under NanoMITe grant (LRGS/2015/UKM-UPM/NanoMITe/04/02). Authors also wish to thank staffs and students of Functional Devices Laboratory and ITMA for helping with the equipment facilities, simulation and fabrication process of the antenna.

6. References

[1] Prudenziati M 1994 *Handbook of Sensors and Actuators 1: Thick Film Sensors* vol 1, ed S Middelhoek (Amsterdam: Elsevier) p 113
[2] Rashidian A, Shafai L, Sobocinski M, Peräntie J, Juuti J and Jantunen H 2015 IEEE Trans. *Microw. Theory Techn.* **63** pp 2720-29
[3] Schloemann E 2000 *J. Magn. Magn. Mater.* **209** pp 15-20
[4] Mishra R X, Pattnaik S S and Das N 1993 *IEEE Trans. Antennas Propag.* **41** pp 230-233
[5] Hasan I H, Hamidon M N, Ismail I, Osman R and Ismail A 2017 *Proc. IEEE Reg. Symp. Micro and Nanoelectronics (RSM)* Batu Ferringhi Penang pp 131-134
[6] Ikonen P M T, Rozanov K N, Osipov A V, Alitalo P and Tretyakov S A 2006 *IEEE Trans. Antennas Propag.* **54** pp 3391-99

IOP Conf. Series: Materials Science and Engineering **380** (2018) 012005 doi:10.1088/1757-899X/380/1/012005

Alternative Spin-On-Glass (SoG) material characterization For Deep-Submicron (<0.35 um) soft reflow fabrication process

F Packeer[1], M A M Zawawi[1], N Z I Hashim[1], N M Noh[1], W M Jubadi[2], and M Missous[3]

[1]School of Electrical and Electronic Engineering, Engineering Campus, Universiti Sains Malaysia, 14300 Nibong Tebal, Pulau Pinang, Malaysia
[2]Department of Electronics, Faculty of Electrical and Electronic Engineering, University Tun Hussein Onn Malaysia, Parit Raja, 86400 Batu Pahat, Johor, Malaysia
[3]School of Electrical and Electronic Engineering, University of Manchester, Sackville Street, Manchester, M13 9PL, United Kingdom

E-mail: fauzi.packeer@usm.my

Abstract. Currently a submicron soft reflow process developed in University of Manchester uses silicon nitride (Si_3N_4) as the hard mask layer to support the T-Gate structure of pHEMTs in order to make it mechanically stable. However, an alternative material known as Spin-on-Glass (SoG) is introduced to replace Si_3N_4, offering shorter processing time and consequently, a much simpler and more cost-effective alternative to the e-beam lithography of nanometre-scale gate length transistors. The SoG deposition through plasma-enhanced chemical vapour deposition process requires only 30 minutes to complete, as opposed to the one-day process of depositing Si_3N_4. In this study, the SoG material used is Silicafilm, and the minimum deposition thickness achieved is 138 nm, enabling 150-nm gate length devices to be fabricated. The SoG is also successfully etched at very low power and pressure (20 W and < 25 mTorr respectively), eliminating the detrimental effect of high power plasma etching to the 2DEG carriers. In addition to that, no film cracks observed even by using a single coating and a single baking temperature of 200 °C. All these results indicate the potential of SoG as a suitable hard mask layer alternative to the silicon nitride for the use soft reflow fabrication process.

1. Overview

Currently a submicron soft reflow process developed in University of Manchester uses Silicon Nitride (Si_3N_4) as the hard mask layer to support the T-Gate structure of pHEMTs in order to make it mechanically stable [1]. However development and characterization of alternative material know as Spin-on-Glass (SoG) is being performed to replace silicon nitride (Si_3N_4) as the hard mask layer giving altogether a truly simple and cost effective alternative to e-beam lithography of nanometre scale gate length transistors [2-4] (refer figure 1a and 1b).

In this study, the minimum thickness that could be achieve with Silicafilm is 138 nm enabling ~ 150 nm gate length devices to be fabricated. The challenging issue of SoG is the etching whereby the required high power (>100 W) plasma etching is detrimental to the 2DEG carriers. Here we demonstrate that the SoG is etchable at very low power (20 W) and very low pressure (< 50 mTorr). Furthermore, most literature [5, 6] use multiple coating and ramping for the baking temperature roughly from 80 °C to 230 °C to avoid film cracks but in this study only single coating and single baking temperature (200 °C) applied, with no cracks observed. This simplified the process even further.

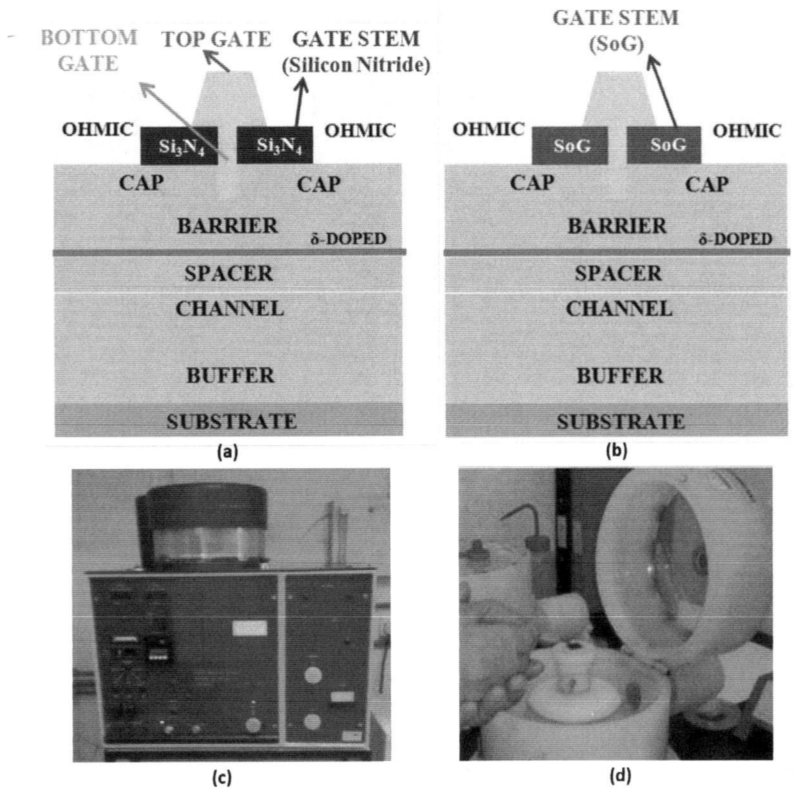

Figure 1. Silicon Nitride (Si_3N_4) (a & c) and Spin-on-Glass (SoG) (b & d) hard mask deposition with its respective equipment.

2. Introduction of SoG

The motivation thus to evaluate SoG material has been of interest to replace Silicon Nitride (Si_3N_4) deposition through Plasma-enhanced Chemical Vapour Deposition (PECVD) by a much simpler and shorter processing time, that is from almost a one day task to only 30 minutes (refer figure 1c and 1d). Basically the technique draws from the planarizing nature of a liquid when applied to a surface topography. The focus is to make the material as close to Silicon Dioxide (SiO_2) as possible. The dielectric and optical characteristics of SoG are quite similar to the characteristics of pure SiO_2 formed by the oxidation of silicon at high temperature.

The SoG material used in this study (Silicafilm) [7], is an alcohol solution which is applied to a semiconductor surface to yield a pure SiO_2 film. Such film may be formed on a wide variety of flat surfaces by spinning, spraying or dipping. However, for precise control of flatness and greatest uniformity, it is performed using spinning methods (refer figure 1d).

3. SoG deposition

Coated sample is spun at the desired speed for thickness of choice for 30 s. During this time, the liquid film will spread uniformly on the entire sample and the solvents in the liquid will eventually evaporate, leaving a thin film of solid component [8] on the sample. Finally, baking the sample on hot plate at

200 °C for 15 minutes will turn the thin film of solid component to SoG (refer figure 2). Some studies [5, 6] use sequences of hot plate bakes from low to high temperature, with one minute baking for each temperature, then high temperature (> 350 °C) curing in an oven with Nitrogen flow. This to avoid flakes.

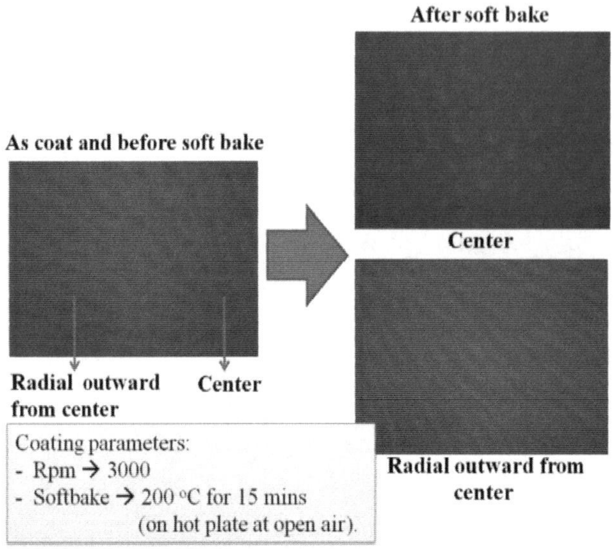

Figure 2. Silicafilm SoG surface profile before and after soft bake.

4. SoG thickness check

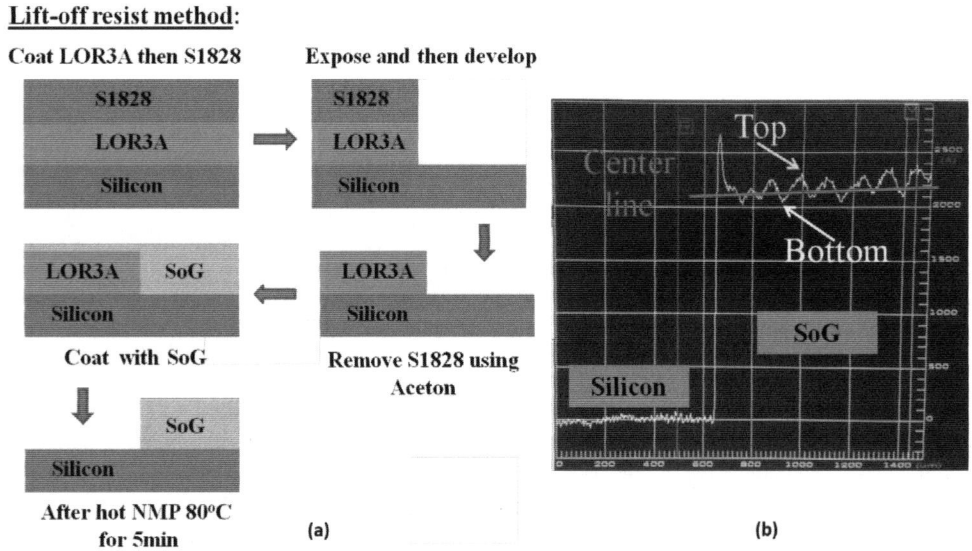

Figure 3. Lift-off process to determine Silicafilm SoG thickness and its measurement at spin-speed 3000 rpm and spin time 30 s.

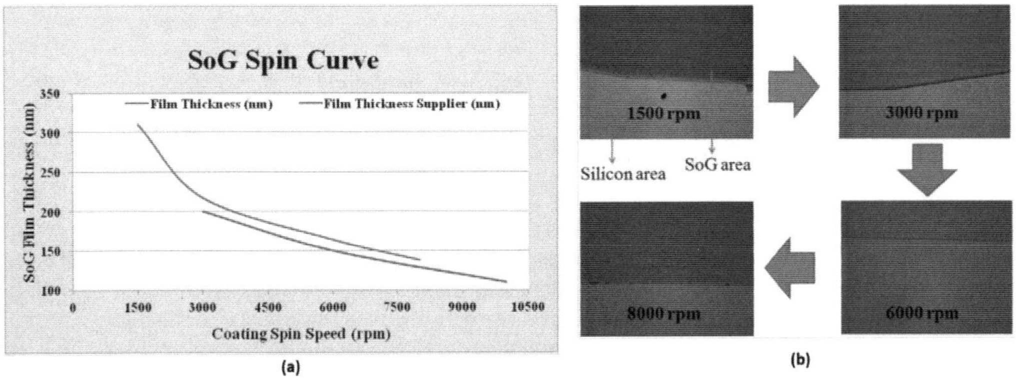

Figure 4. Silicafilm SoG spin-speed curve: in-house data vs. supplier data under same soft-bake conditions (a) and Silicafilm SoG colours at different thickness with respect to spin speed (b).

The thickness check is done using a liff-off process (refer figure 3a). SoG thickness measurement is then performed using a DEKTAK 3 ST surface profiler. The measurement look rough but it is within +/- 5 % variation from the centre value as shown in figure 3b. This lift-off process concept is unique, repeatable and gives a high degree of confidence in the measurement results and much simpler compare to plasma etching.
Figure 4a shows that the thickness of the SoG film will shrink if the spin-speed increases. The thinnest SoG film is at maximum speed i.e. 8000 rpm while the thicker film is at the slowest speed of 1500 rpm. The red curve is the data from the supplier where the minimum spin-speed is 3000 rpm while the maximum spin-speed is at 10000 rpm. The colour of the SoG film is also different from one thickness to another as shown in figure 4b. These colours represent variations in thickness as the light refracts.

5. Pin holes verification

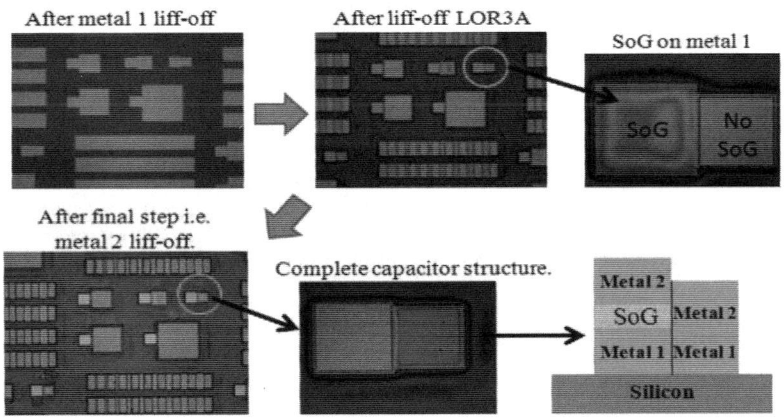

Figure 5. Lift-off process to fabricate MIM capacitor using Silicafilm SoG as dielectric for pin holes verification.

The quality of a thin film is sometimes affected by the defect on the film surface or pinholes, which have potential for causing reliability problem and shorts. Visual inspection shows no pinholes but investigated further by fabricating Metal-Insulator-Metal (MIM) capacitor (figure 5). The concept is, if the thin film which is the dielectric or insulator of the capacitor has pinholes then it will short. Testing for continuity basically means testing to see if there is an electric connection between two points. If two points are electrically connected they are said to be continuous or shorted. In this case it means, if the SoG has pinholes, it will allow metal 2 to be in contact with metal 1 below. Hence, 100 % continuity check on the capacitors give 100 % yield on no short, which means the 216 nm and 138 nm SoG are reliable and suitable to be an alternative hard mask layer for submicron reflow process.

6. SoG etch study

In this study, the etching is done using the Reactive Ion Etching (RIE) plasma chamber. The etching parameters are to be as close as possible to those used to etch Si_3N_4 as previously mentioned i.e. low RF power and low pressure. Low RF power need to be maintained to avoid degradation of the 2DEG later on when making a device and low pressure can minimize loading effect. The etching time must also be reasonably short as longer ones will erode the S1805 resist which is just 500 nm in thickness. All of this restrictions are basically a challenge to etch SoG as reported in most literature [9, 10], to etch SiO_2 or SoG, the minimum RF power used is 100 Watt and the resist thickness is usually more than 2 μm. Figure 6a and 6b show the etch rate and selectivity respectively. While figure 6c illustrate the etch rate on patterned sample.

7. Soft reflow on SoG

The main motivation for the use of SoG is to have alternatives for hard mask layer for the submicron soft reflow study. This SoG coating process is simpler and less time consuming as mentioned previously compared to the Si_3N_4 deposition process. Hence, with the success in the etch rate studies using thicker resists, the study of 138 nm SoG was continued for soft reflow. The trial runs used thick S1805 (with 20 minutes EBR) on patterned samples.

The results for the 90 s soft reflow run are shown in figure 7 on thicker S1805 (with 20 minutes EBR). Base on the result shown figure 7, the soft reflow works on the 138 nm SoG as hard mask layer. The initial 1 μm features shrink after 90 s reflow at a solvent temperature less than 50 °C. Further optimization on soft reflow conditions and etching could enable gate length ~ 150 nm.

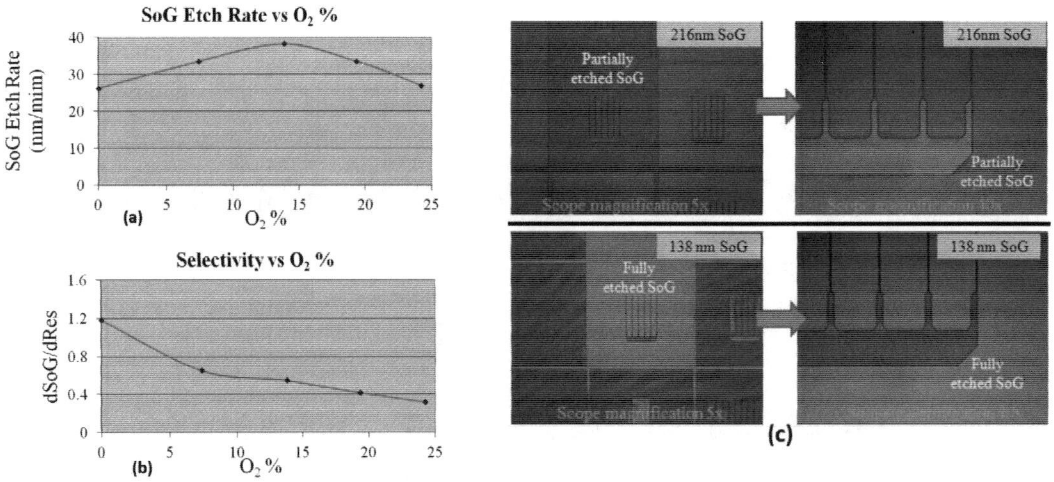

Figure 6. Silicafilm SoG etch rate (a), selectivity (b) and Silicafilm SoG on device pattern after etch & clean (c).

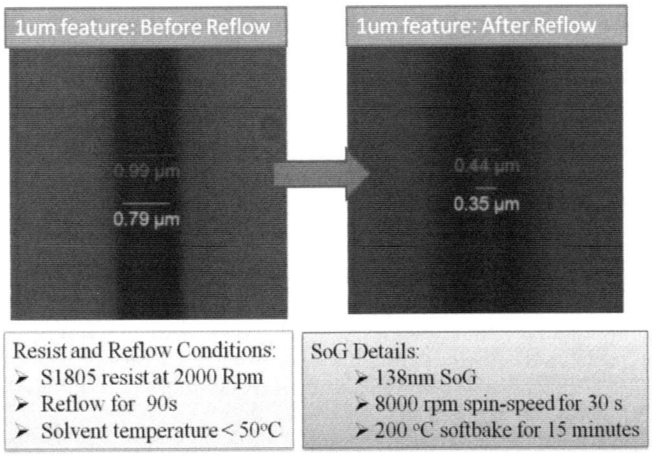

Figure 7. Soft reflow on resist with Silicafilm SoG as hard mask.

8. Conclusions

The possibility of using Spin-on-Glass (SoG) material as an alternative to Silicon Nitride (Si_3N_4) hard mask deposition was explored. The motivation for this study is that the SoG process is much simpler than the Si_3N_4 one and save a lot of processing time. However, the key challenge of this SoG is etching as it needs high power (>100 W). In order to maintain low power (20 W) to avoid degrading the device 2DEG, etch studies were performed and a successful recipe to etch SoG of thickness 216 nm and 138 nm was developed. The developed etching recipe however has low selectivity to resist and attacks the resist around the reflowed area. The future works would be further optimizing the current process of soft-reflow on SoG to realize smaller gate length down to 150 nm, these include using thinner SoG material and using more aggressive etching gas such as Sulfur Hexafluoride (SF_6) or Buffered Oxide Etch (BOE).

9. Acknowledgement

The authors would like to thanks The University of Manchester, United Kingdom for the lab facilities to conduct the experiment and characterization. This work was also partly supported by Universiti Sains Malaysia under Research University Short Term Grant No. 304/PELECT/6315067.

10. References

[1] K. W. Ian, M. Exarchos, and M. Missous, "Fabrication of 250-nm T-gate $In_{.52}Al_{.48}As$-$In_{.70}Ga_{.30}As$ pHEMT using a novel solvent reflow technique at room temperature," in *2012 Ninth International Conference on Advanced Semiconductor Devices & Microsystems (ASDAM),*, 2012, pp. 99-102.

[2] P. C. Chao, P. M. Smith, S. Wanuga, W. H. Perkins, and E. D. Wolf, "Channel-Length Effects in Quarter-Micrometer Gate-Length GaAs MESFET's," *IEEE Electron Device Letters,*, vol. 4, pp. 326-328, 1983.

[3] M. Matsumura, K. Tsutsui, and Y. Naruke, "Submicrometre Lift-Off Line With T-Shaped Cross-Sectional Form," *Electronics Letters,* vol. 17, pp. 429-430, 1981.

[4] P. C. Chao, W. H. Ku, P. M. Smith, and W. H. Perkins, "0.2 Micron Length T-Shaped Gate Fabrication Using Angle Evaporation," *IEEE Electron Device Letters,*, vol. 4, pp. 122-124, 1983.

[5] S. W. Tan, W. T. Chen, M. Y. Chu, and W. S. Lour, "Sub-0.5-µm gate doped-channel field-effect transistors with HEMT-like channel using thermally reflowed photoresist and spin-on glass," *Semiconductor Science and Technology,* vol. 19, p. 167, 2004.

IOP Conf. Series: Materials Science and Engineering **380** (2018) 012005 doi:10.1088/1757-899X/380/1/012005

[6] A. C. Madayag and Z. Zhiping, "Optimization of Spin-on-Glass Process for Multilevel Metal Interconnects," in *Proceedings of the Fourteenth Biennial University/Government/Industry Microelectronics Symposium,*, 2001, pp. 136-139.
[7] EMULSITONE and COMPANY. (12 Nov). *APPLICATIONS OF SILICAFILMS*. Available: http://www.emulsitone.com/sif.html
[8] S. Hirasawa, Y. Saito, H. Nezu, N. Ohashi, and P. Maruyama, "Analysis of Drying Shrinkage and Flow Due to Surface Tension of Spin-Coated Films on Topographic Substrates," *IEEE Transactions on Semiconductor Manufacturing,*, vol. 10, pp. 438-444, 1997.
[9] B. E. E. Kastenmeier, P. J. Matsuo, J. J. Beulens, and G. S. Oehrlein, "Chemical dry etching of silicon nitride and silicon dioxide using $CF_4/O_2/N_2$ gas mixtures," *Journal of Vacuum Science & Technology A: Vacuum, Surfaces, and Films,* vol. 14, pp. 2802-2813, 1996.
[10] S. A. Rosli, A. A. Aziz, and H. A. Hamid, "Highly Chemical Reactive Ion Etching of Silicon in CF_4 Containing Plasmas," in *IEEE International Conference on Semiconductor Electronics, 2006. ICSE '06.* , 2006, pp. 856-860.

IOP Conf. Series: Materials Science and Engineering **380** (2018) 012006 doi:10.1088/1757-899X/380/1/012006

The Effect of Indentation Load, Tilted Angle, and the Amount of Surface Removal to the Fracture Toughness of Silicon Nitride Measured by the Surface Crack in Flexure (SCF) Test Method

T. G. T. Nindhia[1]

1. Department of Mechanical Engineering, Engineering faculty, Udayana University, Jimbaran, Bali, Indonesia
E-mail: nindhia@yahoo.com/tirta.nindhia@me.unud.ac.id

Abstract. The surface crack in flexure (SCF) is a method for the evaluation of the fracture toughness of advanced ceramics. Conventionally is practiced by using a Knoop indenter to make a very small precrack. Removal on indent and the plastically deformed zone is required before the fracture test. The purpose of this removal is to eliminate residual stresses under the Knoop impression and to obtain a semi elliptical precrack shape. In this work the influence of the variation of several experimental parameters is investigated. Fracture toughness values by the SCF method are compared with those measured using the SEVNB (single edge V-notched beam) method. The material chosen for this purpose was gas pressured sintered silicon nitride (Si_3N_4) containing 3wt.%Al_2O_3 and 3 wt.%Y_2O_3 (SL200B, Ceram Tec, Plochingen, Germany). The varied parameters were indentation load, orientation of the indentation crack with respect to the bending axis and the amount of material removed from the surface. Additional investigations were performed to determine the crack geometry for various indentation loads. A procedure was developed to facilitate the location and measurement of the precrack size on the fracture surfaces. The effect of measurement accuracy of the results is evaluated. The fracture toughness of specimens with a surface removal in the range suggested from ASTM C 1421 were found to agree with the results obtained from SEVNB. A surface removal below the recommendation resulted in low values of fracture toughness. Increasing the amount of surface removal moderately was found to still fit with results obtained from SEVNB. Surface removal of much more from the recommended amount leads to failure from natural flaws. Observations from serial sectioning revealed that precracks obtained from HK20 and HK30 have an irregular precrack shape with large lateral cracks. It is suggested to use HK5 and HK10 to produce semi-elliptical surface cracks.

1. Introduction

The surface crack in flexure (SCF) is a method for the evaluation of the fracture toughness of advanced ceramics also known as controlled flaw method. Despite this good track record, the SCF method still has the drawbacks. Previous research informed that with serial-sectioning experiments beneath Knoop indentations, observed that a second set of lateral cracks developed very deep beneath the indentation at indentation loads above 98 N [1]. Quinn and Lloyd [2] also detected deeper than expected lateral cracks in a particular machinable glass-ceramic specimen. The oversized lateral crack

IOP Conf. Series: Materials Science and Engineering **380** (2018) 012006 doi:10.1088/1757-899X/380/1/012006

was estimated to have interfered with the median crack to the extent that fracture toughness [3,4]. To overcome the lateral crack problem Quinn [4] suggested the lateral cracks may be eliminated by hand grinding or polishing off 7–10 times the indentation depth. The requisite amount can be checked by monitoring the hand-ground tensile surface before specimen fracture. This technique is not convenient to conduct since we need to check regularly in the time of grinding or polishing. Regarding determination the indentation load for the precrack, the standard C 1421 only inform that the determination basic on class of materials *i.e.* approximately 10 to 20 N are suitable for very brittle ceramic, 25 to 50 N for medium tough ceramic, and 50 to 100 N for very tough ceramic[5]. This procedure is not having any significant related with lateral crack problem.

It was reported on previous publications [6,7,8] for silicon nitride in some circumstances, the determination of precrack size was difficult to detect. Park e*t al.* investigated the effect of crosshead speed on the fracture toughness of Si_3N_4. A slow crosshead speed was applied in order to yield slow crack growth (SCG) resulting "halo" region to appear. With appearance of halo region the measurement of the precrack size will be easier to conduct. Unfortunately the halo region induced by SCG did not appear in the Si_3N_4 due to critical crack size at fracture is equal to the initial precrack size induced by indentation [9].

Serial sectioning is introduced in this research to overcome the problem of lateral crack and difficulties in determination of precrack size. It is included also in this work to investigate the influence of the variation of several experimental parameters on fracture toughness by using the SCF method. The results are compared with those measured using the SEVNB (single edge V-notched beam) method [10].

2. Experimental

The material chosen was silicon nitride produced by Ceram Tec (Plochingen, Germany) under the name SL200 B (Table 1). It is a gas pressure sintered ceramic containing ~3 wt.% Al_2O_3 and ~3 wt.% Y_2O_3. This material was a silicon nitride reference material [11]

Table 1. Material Properties

Density	3190±9 kg/m^3
Young modulus	303± 1.3 (bulk), 307± 2(skin)GPa
Hardness HV1, HV 5, HV10	1452, 1461, 1446

SEVNB, the fracture toughness was conducted based on preparation of bar test pieces in which a sharp-tipped notch is machined. The reciprocating razor blade and diamond paste was used for honing the test piece. Under well controlled condition, a notch-tip radius in the range of 1 μm to 20 μm was be obtained for valid condition regarding the standard. The bar test pieces of rectangular with cross section size of 3 mm x 4 mm and length of ≥ 4.5 mm was prepared for these purpose. The test piece in the jig on its 3 mm width face with the V-notch face resting on the outer-support roller. The four-point flexure was employed in this research. The loading rate was applied at 0.5 mm/min. The depth of V-notch was measured by observing the fracture surface using a microscope. Fracture toughness was computed in accordance with ISO/FDIS 23146 and the value was compared to the one that obtained from SCF method.

SCF, The bend specimen for this purpose was prepared with cross section size of 3 mm x 4 mm and length of ≥ 4.5 mm, or halves of previously tested strength specimens. In order to obtain fine precrack, preliminary research with serial sectioning was carried out on 4 variations of Knoop indentation load namely: 5 kg, 10 kg, 20 kg, and 30 kg. Knoop indentation loads that yield similar precracks with small lateral crack were chosen. The specimen then was precracked with 2 variations of tilted angles, that are 0.5° and 1° in order to observe in which angle would be easier to discern and also to investigate the effect of tilted angle to the value of fracture toughness. After precracking, the residual stress and lateral crack under the indentation impression must be removed. The standard

IOP Conf. Series: Materials Science and Engineering **380** (2018) 012006 doi:10.1088/1757-899X/380/1/012006

C1421 introduced the amount removed should be 0.150d-0.167d which is d is length of the long diagonal of Knoop impression and the removing process was done by using hand grinding. The four-point flexure was employed in this research. The loading rate was applied at 1 mm/min. Fracture toughness was computed in accordance with ASTM C 1421. It is also included in this research to investigate the effect surface removed on fracture toughness below and above suggested by the standard.

3. Result and Discussion

The detail of the precracks obtained from different indentation loads is presented in Fig. 1. The suitable precracks were obtained from Hardness Knoop with indentation load 5 kg (HK5) and with load 10 kg (HK10) as can be seen at Fig. 1a and 1b. The precracks were found contain only small size of damage zone and short-shallow lateral crack, therefore only small amount of surface removal needed. In contrary with load 20 kg (HK20) and 30 kg (HK30), the precracks were found irregular, contain large damage zone with long and deep lateral cracks. Some of the precraks in these loads were found with second deep lateral cracks. Since lateral cracks may interfere with the primary median crack and cause errors in determination of fracture toughness [1,4], and with 0.150d - 0.167d will not be sufficient to completely remove the lateral crack, then HK20 and HK30 were not used in this research. Result of precrack size obtained from serial section also useful to facilitate the location and measurement of the precrack size on the fracture surfaces by fitting it in the fracture surface.

Figure 2 shows the effect of fracture toughness, as a function of material removal to be compared with one that obtained from SEVNB method. The fracture toughness of specimens with a surface removal in the range suggested from ASTM C 1421 was found to agree with the results obtained from SEVNB. A surface removal below the recommendation resulted in low values of fracture toughness. Increasing the amount of surface removal moderately was found to still fit with results obtained from SEVNB. Surface removal of much more from the recommended amount leads to failure from natural flaws.

To recognise whether the failure is caused by the precrack or by natural flaws, one can observe the crack propagation after specimen broken. The specimen will broke in to three pieces if failure caused by natural flaws as can be shown in Fig. 2 . It is also possible to be done by observation of fracture surface. If the specimen failure caused by the precrack the fractography on the fracture surface will appear with quite large size fracture mirror.

The influent of several parameters on fracture toughness by using SFC method to be compared with SEVNB method can be observed on Fig. 3. The fracture toughness for SCF method reach the peak value when amount of surface removal approaching the maximum amount surface removal designated by the standard (0.167d). From this peak value, if amount of surface removal was increased, the values of fracture toughness were found decrease. In general, the fracture toughness value obtained with 0.5^{o} tilted angle resulting higher values comparing with one that have tilted angle 1°.

It is still found recently that fracture toughness tests for silicon nitride [12,13,14] were conducted by using indentation fracture test method which is not appropriate method since the indentation method is non standardized method for fracture toughness test. It is suggested to use improved SCF standardized method that is introduced in this article

IOP Conf. Series: Materials Science and Engineering **380** (2018) 012006 doi:10.1088/1757-899X/380/1/012006

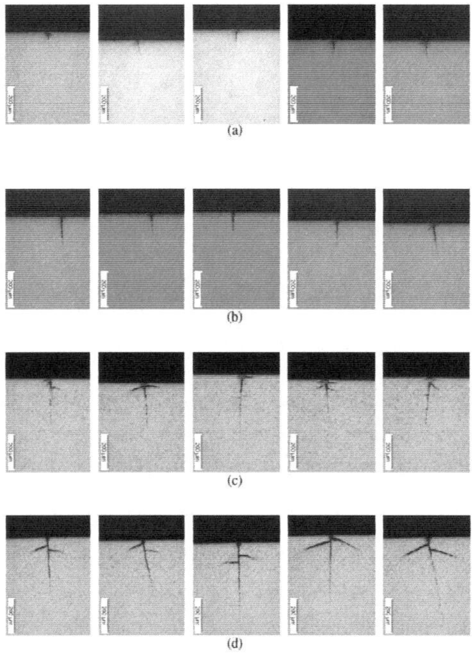

Figure 1. Typical of suitable precracks that contain small size of damage zone with short lateral cracks were obtained from HK5(a) and HK10(b). An irregular form of lateral crack and second deep lateral crack arise from HK 20 (c). A long and deep lateral crack appeared from HK30 (d).

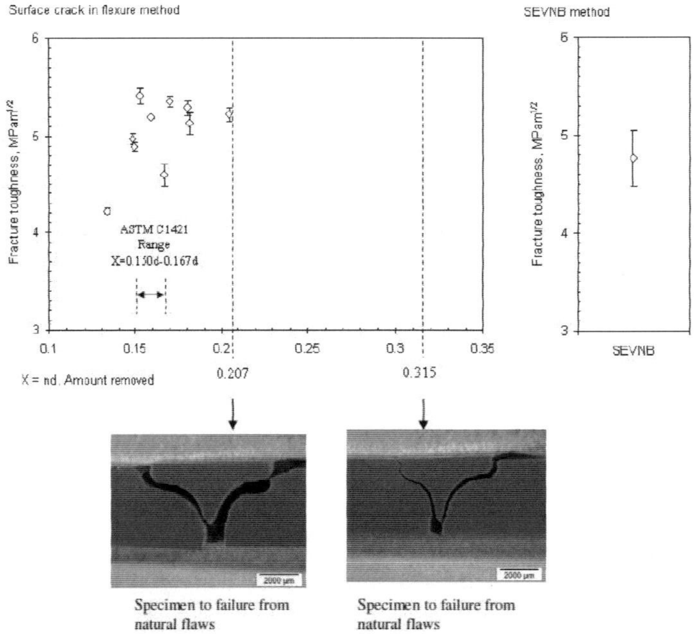

Figure 2. The effect of amount surface removed on fracture toughness in SCF method to be compared with SEVNB methods. Surface removal of much more (0.207d and 0.315d) from the recommended amount leads to failure from natural flaws

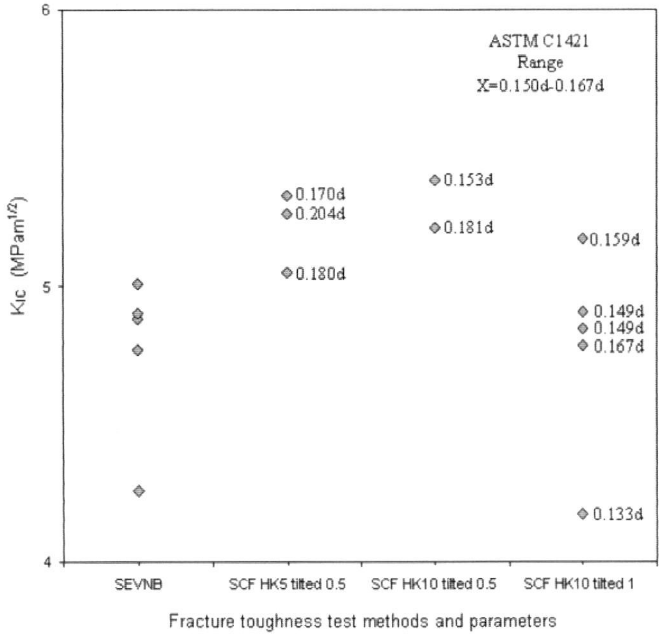

Figure 3. The effect of indentation loads, tilted angles, and amount surface removed on fracture toughness

4. Conclusion

By Serial section is useful to facilitate the location and measurement of the precrack size on the fracture surfaces. Other benefits obtained from serial sectioning is the amount of surface removing also can be determined. From precrack shape that obtained from serial sectioning a suitable indentation load can be selected to avoid irregular or large lateral cracks that may interfere the fracture toughness value.

The fracture toughness of specimens with a surface removal in the range suggested from ASTM C 1421 was found to agree with the results obtained from SEVNB. A surface removal below the recommendation resulted in low values of fracture toughness. Increasing the amount of surface removal moderately is found to still fit with results obtained from SEVNB. Surface removal of much more from the recommended amount leads to failure from natural flaws. The fracture toughness value obtained with 0.5° tilted angle for the SCF method resulting higher values comparing with one that have tilted angle 1°.

5. Acknowledgements

The author acknowledge Prof. Dr. Robert Danzer and Assistance Professor Dr. Tanja Lube of *Institut für Struktur und Funktion Keramik, Montanuniversität*, Leoben, Austria on supervising and facilitating this research and the ASEA-UNINET for providing travel and accommodation. The cost for publication is supported by Engineering Faculty, Udayana University, Jimbaran, Bali, Indonesia.

References

[1] Lube T., 2001, Indentation crack profiles in silicon nitride, *Journal of The European Ceramic Society*. 21, pp.211-218.

[2] Quinn JB, Lloyd IK., 2000, Comparison of methods to determine the fracture toughness of three glass-ceramics at elevated temperature, *J. Am. Ceram. Soc.*, 83, pp. 3070–76.

[3] Quinn GD, Swab JJ, Hill MD., 1997, Fracture toughness by the surface crack in flexure (SCF) method: new test results, *Ceram. Eng. Sci. Proc.,* 18, pp. 163–72.

[4] Quinn GD, Salem JA., 2002, Effect of lateral cracks on fracture toughness determined by the surface-crack-in-flexure method, *J. Am. Ceram. Soc.*, 85, pp. 873-80.

[5] ASTM C1421-99, Standard test method for determination toughness of advanced ceramic at ambient temperature.

[6] Quinn GD, Swab JJ., 2003, Fracture toughness of a toughened silicon nitride by ASTM C 1421, *J. Am. Ceram Soc.* 86, pp.1043-45.

[7] Getting RJ, Quinn GD., 1995, Surface crack in flexure (SCF) measurement of the fracture toughness of advanced ceramic, *Ceram. Eng. Sci.* Proc. 16, pp. 539-47.

[8] Quinn GD, Getting RJ, Kübler JJ., 1996, Fractography and surface crack in flexure (SCF) method for evaluating fracture toughness of ceramic, pp.107-44, ceramic transaction, Vol. 64, Fractography of glass and ceramic III, Edited by Varner JP, Frech VD, Quinn GD, American Ceramic Society, Westerville, OH.

[9] Park JK, Yasuda K, Matsuo Y, 2001, Effect of crosshead speed on the fracture toughness of soda lime glass, Al_2O_3, and Si_3N_4 ceramics determined by the surface crack in flexure (SCF) method. *Journal of Material Science*, 36, pp. 2335-2342.

[10] ISO/FDIS 23146:2008(E), Fine ceramic (advanced ceramic, advanced technical ceramics)-test methods for fracture toughness of monolithic ceramic-single edge notch beam (SEVNB) method.

[11] Lube T, Dusza J., 2007, A silicon nitride reference material-A testing program of ESIS TC6, *Journal of the European Ceramic Society*. 27 , pp. 1203–1209.

[12] Nishiyama, N. Ishikawa, R. Ohfuji, H, Marquardt, H., Kurnosov, A., Taniguchi, T., Nam Kim, B., Yoshida, H., Masuno, A., Bednarcik, J., Kulik, E., Ikuhara, Y., Wakai, F., and Irifune, T., Transparent polycrystalline cubic silicon nitride. *Sci. Rep.* **7**, 44755, 2017, pp.1-8.

[13] Kvetková , L., Duszová , A., Kašiarová , M., Dorčáková , F.,Dusza , J., and Balázsi, C, Influence of processing on fracture toughness of Si_3N_4 + graphene platelet composites, *Journal of the European Ceramic Society*, 33, 2013, pp. 2299-2304

[14] Wang, L., Qi , Q., Cai, P., Zhang, H., Yang, X., Liu, X., Jiao, Z., and Huang, Z., New route to improve the fracture toughness and flexural strength of Si_3N_4 ceramics by adding $FeSi_2$, *Scripta Materialia*, 126, 2017, pp. 11–14

IOP Conf. Series: Materials Science and Engineering **380** (2018) 012007 doi:10.1088/1757-899X/380/1/012007

Influence of uniaxial deformation on structure of carbon-coated polyurethane surface

I A Morozov[1,2], A S Mamaev[3], A Y Beliaev[1] and I V Osorgina[2]

[1] Institute of continuous media mechanics UB RAS, Academika Koroleva st. 1, 614013, Perm, Russia
[2] Perm State University, Bukireva st. 15, 614990, Perm, Russia
[3] Institute of Electrophysics UB RAS, Amundsen st. 106, 620016, Ekaterinburg, Russia

Email: ilya.morozov@gmail.com

Abstract. Plasma treatment of polymers is the promising technique for improving the quality of biomedical devices, in particular deformable implants. The influence of mechanical deformation on such coating has not been sufficiently studied. The effect of the uniaxial cyclic deformation on the surface structure of carbon nanolayer deposited (by the magnetron sputtering) on soft elastic polyurethane is investigated. The obtained coating is wrinkled. The mechanical load changes the structure of the surface (wavelength, amplitude and fractal dimension of the wrinkles, surface roughness); the cracks appear on the deposited layer. As the strain increases, the number of cracks rises and the strain-induced folds appear on the coating.

1. Introduction

Plasma treatment of materials improves various characteristics of products. To achieve certain goals (increase of hardness, wear and corrosion resistance [1], adhesion, biomedical characteristics [2-4]) various types of plasma treatment (sputtering, implantation, etching), environmental conditions (working gas, sputtering material), experimental properties (duration, temperature, frequency of pulses, etc.) are applied.

The effect of plasma on hard materials (metals, ceramics, plastics) has been studied extensively. Such treatment changes the chemical structure, relief and energy of the surface. In particular, modification of polymer surface [5] improves biomedical parameters: reduction of thrombogenesis [6], biocompatibility with cells and body tissues [7]. This feature opens a promising challenge of creating deformable biomedical implants with improved properties. These benefits are related to the surface energy [8], hardness [9], presence of free radicals [10] and the relief structure: unlike hard materials, the plasma treatment of soft materials leads to the formation of a wrinkled surface [11, 12]. The texture of the relief has a certain effect on the adhesion of bacteria [13] and the wrinkles of the surface significantly reduce the adhesion of staphylococcus [14].

Under the real-life conditions, elastomeric materials undergo mechanical deformations. What happens to the plasma modified surface exposed to external loading? Qualitative studies of stretched plasma-treated polymer films show regular structure of cracks and folds [14, 15]. How this will affect the interaction with the environment? To date, these issues have been studied extremely insufficient.

In this paper, the effect of cyclic uniaxial loading on the surface structure of soft elastic polyurethane coated with a carbon nanolayer deposited by the magnetron sputtering method were investigated. The changes in the surface structure after loading, the formation of cracks and strain-induced folds were shown.

2. Materials and methods

Preparation of samples. Polyurethane was made from the two-component system of prepolymer (urethane based polyether and toluene diisocyanate) and hardener (solution of aromatic diamine in polyol); the weight ratio was 100:46.2. Prepolymer was heated to 50 °C and mixed with the hardener. The mixture was press-molded and cured at 100 °C for 20 hours. The thickness of the plates was 2 mm. Plasma treatment of the materials was carried out by the magnetron sputtering of graphite target with high power pulses (HiPIMS – High-Power Impulse Magnetron Sputtering) in Ar medium, the pressure was 0.2 Pa. The voltage of the magnetron discharge was 800 V, the current amplitude – 80 A, the pulse duration and rate – 150 μs and 150 Hz. The temperature of the samples did not exceed 100 °C. This regime is also called mixed HiPIMS-Arc and has a number of advantages: prevention of development of large defects of the deposited coating, high sputtering speed and significant fraction of sp3-bonds.

The sample was placed on an electrically insulated holder at a distance of 10 cm from the target of the magnetron. The deposition time was chosen by the measurement of the resistance R of the coating using mega-ohmmter (APPA 605), which allows to measure resistance up to 22 GΩ. The resistance of the carbon coating deposed for 8 min was 15 GΩ. The deposition less than 8 min resulted in $R > 22$ GΩ. Four samples were prepared with the next duration of treatment: 8 min – (labeled as PU-8m), 6 min – (PU-6m), 4 min – (PU-4m) and 2 min – (PU-2m). The thickness of the coating was 25, 19, 13 and 6 nm respectively (measured by the depth of the scratch of the same layer deposited on a glass substrate).

The treated samples (20x5x2 mm strips) were subjected to cyclic uniaxial loading (dynamic modulus analyzer DMA / SDTA861e, METTLER TOLEDO): 1000 cycles, strain amplitude – 10%, frequency – 1 Hz. Due to the small thickness of the coating, no differences in mechanical properties were found.

Microstructure of the surfaces before and after the mechanical loading was investigated by the atomic force microscope (AFM Dimension Icon). The stiff probes (RTESP) with tip radius of 10 nm were used. For the sake of statistics, from each sample several images 10x10 μm (resolution in xy-plane 1024x1024 dots) were obtained. The microstructure of strained samples was also studied in situ. In this case the sample was fixed and stretched in the small stretching device placed under the AFM scanner.

3. Results and discussion

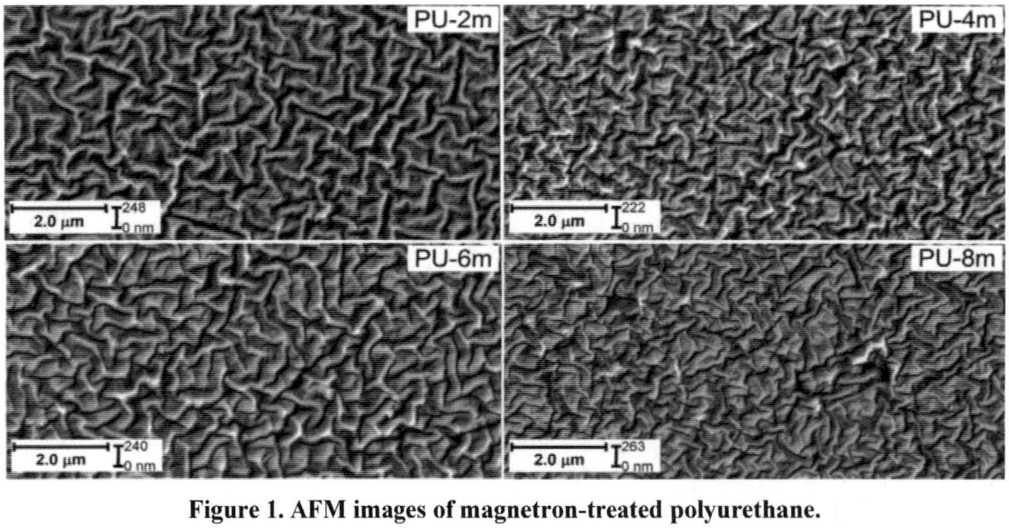

Figure 1. AFM images of magnetron-treated polyurethane.

The surface of the untreated polyurethane (not shown here) has a smooth relief without pronounced features. The surface with the deposited carbon layer has a wrinkled structure (figure 1, hereafter the AFM images are cut vertically for the sake of space). The nature of wrinkling is complex and involves both the temperature mechanism [16] accompanied with the heterogeneous structure of polyurethane at the nanoscale: rigid domains in a soft matrix.

The structural properties of the surface depend on the treatment. For quantitative analysis, wrinkles were identified by the watershed method: the regions above the mean value of height of the image were analyzed.

The surfaces after the mechanic tests are given in figure 2. Qualitative inspection of these images showed that the relief of PU-2m did not visually change; the materials PU-2m-6m has randomly oriented tiny cracks. Pronounced long cracks are visible on the material PU-8m.

The area A and the perimeter P of the contours of the wrinkles in double logarithmic coordinates ($\lg(A)$; $\lg(P)$) form a straight line (coefficient of determination of the linear approximation is $R^2 > 0.95$). Therefore, A and P of the wrinkles has the fractal dependence: $P \sim A^{D/2}$, where D is the fractal dimension.

The wave properties of the two-dimensional relief z were analyzed by the power spectral density: $PSD = L/n^2 \operatorname{abs}(\operatorname{fft}(z))^2$, where fft is the Fourier transform, n is the number of points of the AFM image of side length L. PSD shows the strength of the signal as a function of frequency f. Spatial discrete frequencies are: L^{-1}, $2L^{-1}$, ..., $0.5nL^{-1}$; the wavelength is $\lambda = 1 / f$. The obtained reliefs are isotropic (the two-dimensional PSD has the radial symmetry); therefore in calculations the one-dimensional radial PSDs were used.

The root mean square roughness in the frequency range $f_1 \ldots f_2$ is calculated as:

$$R_{RMS} = \sqrt{\int_{f_1}^{f_2} PSD df}. \tag{1}$$

In particular, if $f_1 = f_2$ the amplitude A for the wavelength $\lambda = 1 / f_1$ is estimated; if $f_2 = \max(f)$ the average roughness in a region of size $s = 1 / f_1$ is calculated.

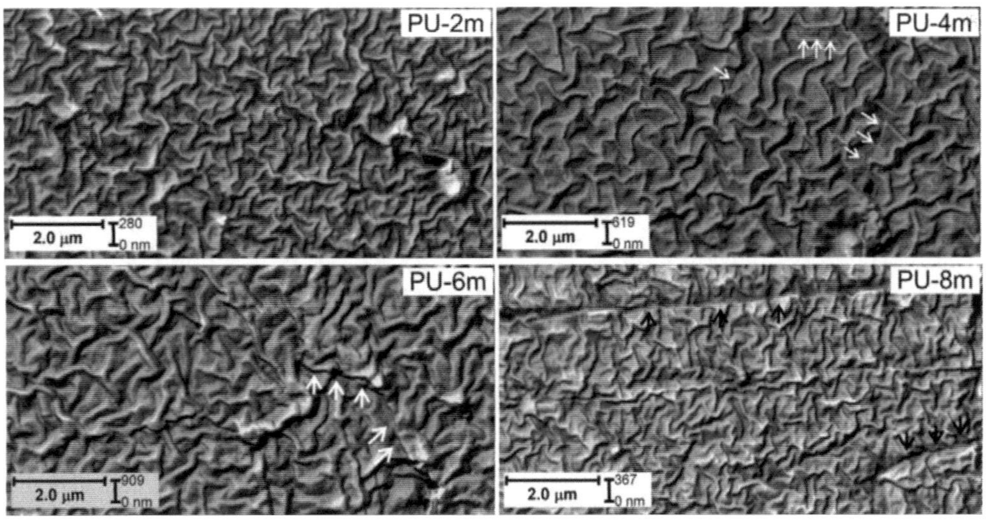

Figure 2. AFM images of the treated surfaces after cyclic uniaxial loading. Axis of deformation is vertical. Boundaries of cracks are marked by arrows.

The quantitative analysis of the surfaces before and after the mechanical load established the changes of the structural parameters of the wrinkled relief.

The roughness vs. size of the region is shown in figure 3(a) (R_{rms} of the untreated material was ~10 nm). As the area of observation increases, R_{rms} rises and at s = 2...3 μm goes to the asymptote. The roughness of materials before the mechanical load changes nonlinearly: it is maximal at t = 6 min, and goes to minimum at t = 8 min. The fractal dimension of the wrinkles insignificantly decreases (figure 3b).

After the mechanical load, roughness of the materials increases. The highest changes, in comparison with the unloaded material, were achieved at t = 4 min; the smallest – 8 min. The fractal dimension of the wrinkles significantly falls as thickness of the coating grows, i.e. wrinkles became more tortuous after the deformation.

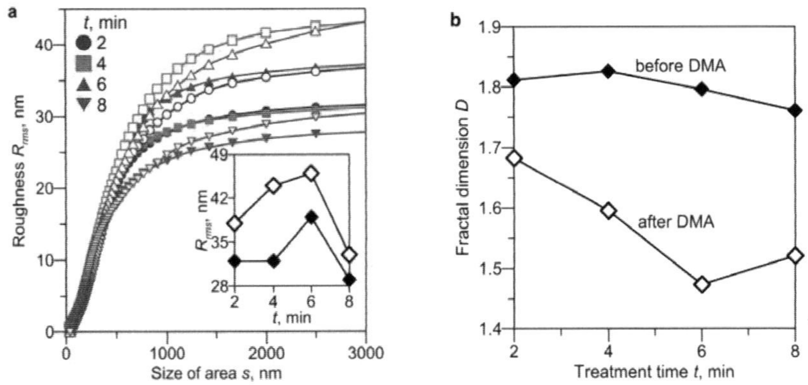

Figure 3. Surface roughness (a) and fractal dimension of the wrinkles (b) of the treated materials before (filled markers) and after (unfilled) the mechanical loading.

The wave characteristics of the surfaces are shown in figure 4. The maxima on the graphs correspond to the amplitudes and wavelengths of the wrinkles. Note, that before the load, there is no clear maximum $A(\lambda)$: the amplitude of wrinkles in the wavelength range of 400…600 nm (t = 2...6 min) and 600…800 nm (8 min) changes insignificantly.

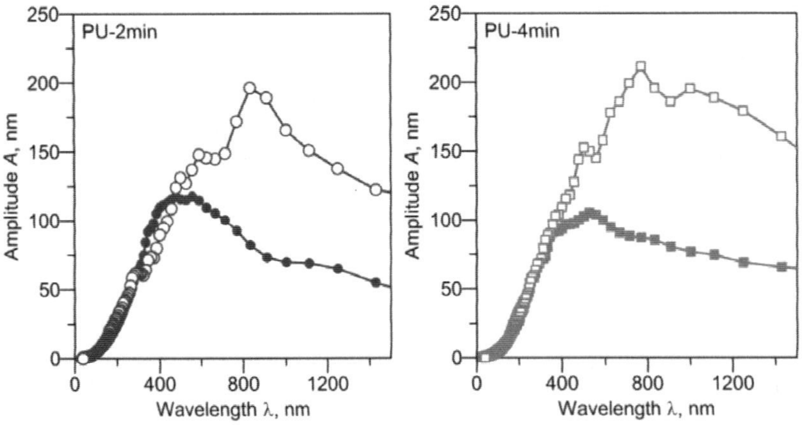

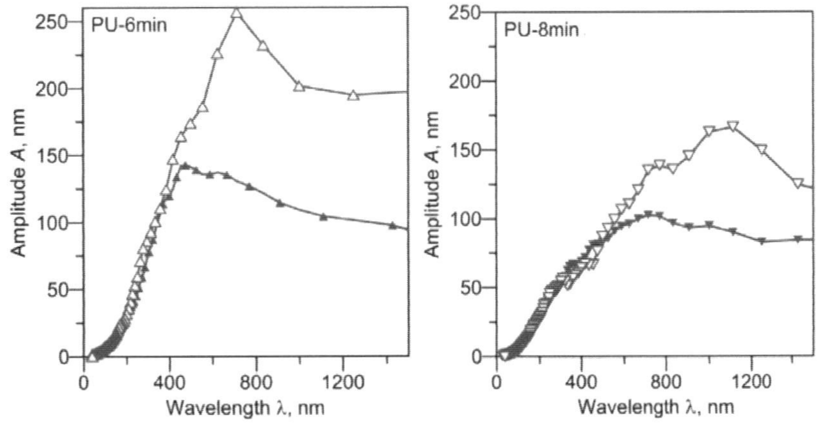

Figure 4. Amplitudes and wavelengths of wrinkles: before (filled markers) and after (unfilled) the mechanical loading.

After the deformation, the amplitude and wavelength of the wrinkles increase – the maxima on the graphs (figure 4) shift to the right, and the peaks of $A(\lambda)$ are sharper, i.e. wrinkles become more pronounced. Moreover, additional local maxima $A(\lambda)$ appear after the mechanical load.

The mechanical load considered above was only 10%. In the case of elastic polyurethane, capable of stretching by hundreds of percent, such deformation is small. Figure 5 shows the relief of PU-2m stretched to 25% and 100%.

Figure 5. AFM images of PU-2m stretched up to 25% and 100%.

The open cracks are visible on the stretched surface. As the load increases, folds, parallel to the loading direction, appear (the result of material compression).

The removal of such high load doesn't restore the initial surface even qualitatively (as in the case of the low load, figure 3), the traces of cracks and longitudinal folds will exist permanently on the unloaded surface.

4. Conclusions

The effect of deformation on the structure of carbon nanolayer deposited by magnetron sputtering on the surface of soft elastic polyurethane was investigated. The thickness of the coating depends on the treatment time (2, 4, 6, 8 min) and was 6, 9, 13, 25 nm. The treated surface has a fractal wrinkled structure, the parameters of which vary nonlinearly with the duration of treatment. Each surface is characterized by a wavelength spectrum of 200 nm width and amplitude from 100 to 150 nm.

The samples were subjected to cyclic uniaxial deformation (strain amplitude 10%, frequency 1 Hz). Mechanical load changed the structure of the surface: the roughness increased, the wrinkles became more tortuous, their amplitudes and wavelengths increased. These structural changes are related to inelastic properties of the coating.

Single microcracks were observed on the coatings of 9…13 nm thickness; in the case of 25 nm extended, orthogonal to the loading axes, cracks appeared. No cracks were found on the 9 nm coating. Investigation of materials in the stretched state established the appearance of open cracks (25% of deformation). At 100% of deformation the cracks are accompanied with folds, caused by the compression of the material.
Changes in the structure of surfaces after loading, as well as damage of the coating, require further study of interaction of plasma-treated deformed polymers with the environment, in particular, with biological tissues. Another possible direction is the development of discontinuous coatings [6] which could be more resistant to the external deformations.

5. Acknowledgement

The work is supported by Russian Science Foundation, Grant №17-79-20042.

6. References

[1] Loganathan A, Ashutosh S, Chris R, Cheng Z, Sara R, Tapas L, Benjamin B and Arvind A 2018 *Surf. Coat. Technol* **334** 384–93.
[2] Liston E M, Martinu L and Wertheimer M R 1993 *J. Adhes. Sci Technol* **7** 1091–127.
[3] Cheruthazhekatt S, Černák M, Slavíček P and Havel J 2010 *J. Appl. Biomed* **8** 55–66.
[4] Hauert R, Thorwarth K and Thorwarth G 2013 *Surf. Coat. Tech* **233** 119–30.
[5] Oehr C 2003 *Nucl. Instrum. Meth. B* **208** 40–7.
[6] Alekhin A P, Boleiko G M, Gudkova S A, Markeev A M, Sigarev A A, Toknova V F, Kirilenko A G, Lapshin R V, Kozlov E N and Tetyukhin D V 2010 *Nanotechnologies in Russia* **5** 696–708.
[7] Slepicka P, Slepickova Kasalkova N, Stranska E, Bacakova L and Svorcik V 2013 *eXpress* Polymer *Letters* **7** 535–45.
[8] Stüber M, Niederberger L, Danneil F, Leiste H, Ulrich S, Welle A, Marin M and Fischer H 2007 *Adv. Eng. Mater* **10** 1114–22.
[9] Tjong S C, and Liu C 2017 *Express Polymer Letters* **11** 434.
[10] Kosobrodova E A, Kondyurin A V, Fisher K, Moeller W, McKenzie D R and Bilek M M M 2012 *Nucl. Instrum. Meth. B* **280** 26–35.
[11] Tsougeni K, Tserepi A, Boulousis G, Constantoudis V and Gogolides E 2007 *Plasma Process. Polym* **4** 398–405.
[12] Croll A B and Crosby A J 2012 *Macromolecules* **45** 4001–6.
[13] Perera-Costa D, Bruque J M, González-Martín M L, Gómez-García A C and Vadillo-Rodríguez V 2014 *Langmuir* **30** 4633–41.
[14] Morozov I A, Mamaev A S, Osorgina I V, Lemkina L M, Korobov V P, Belyaev A Yu, Porozova S E and Sherban M G 2016 *Mat. Sci. Eng. C* **62** 242–48.
[15] Volynskii A L, Nazarov V G, Yarysheva L M, Stolyarov V P, Tsypysheva S V, Bagrov D V, Rukhlya E G, Yarysheva A Yu and Bakeev N F 2012 *Doklady Physical Chemistry* **442** 5–7.
[16] Morozov I A, Mamaev A S, Osorgina I V, Beliaev A Y, Izumov R I, and Oschepkova T E 2018 *J. Appl. Polym. Sci* DOI: 10.1002/app.45983.

IOP Conf. Series: Materials Science and Engineering **380** (2018) 012008 doi:10.1088/1757-899X/380/1/012008

Effect of carbon-based hybrid nano-fillers on the thermal conductivity of PCM composites

U Shahid[1], S Ahzi[1] and A Abdala[2]

[1] Department of Sustainable Development, Hamad Bin Khalifa University, Doha, Qatar
[2] Department of Chemical Engineering, Texas A&M University at Qatar, Doha, Qatar

Abstract. PCMs show great promise as thermal energy storage (TES) medium; however, their low thermal conductivity presents a major bottleneck for their potential application. Enhancement of the thermal conductivity of a paraffin based PCM material using GNP in combination with CNF, CNTs and TRG was investigated in this work. SEM was used to qualitatively assess the dispersion & distribution of the hybrid-nanofillers. DSC was used to determine the melting temperature, thermal capacity and latent heat, whereas thermal conductivity was measured using Hot Disk TPS Thermal Conductivity Instrument (TPS2500S). An overall increase by 143%, 158% and 174% in thermal conductivity and 179%, 193% and 214% in thermal diffusivity was observed for 2 wt.% hybrid loading of TRG, CNT and CNF respectively (pure PCM as ref.). Furthermore, the contribution of the CNF and CNT hybrid-fillers was evaluated to increase the composite PCM thermal conductivity by 122% & 110%, and diffusivity by 117% & 105% respectively (10%-GNP as ref.).

1. Introduction

Efficient energy storage and transfer has been perhaps the most challenging hurdle in the energy industry at the moment. The need to conserve energy and improve its utilization is increasing day by day with the increase in energy demand and growing notion of sustainability. The focus more recently, especially in Qatar, has been on sustainable alternates (i.e. Solar) to meet the local energy demand[1]. Phase change material (PCM) based latent heat storage systems have emerged as a very effective technique for thermal management systems[2] and have gathered significant interest in applications like solar energy systems, thermal energy storage (TES), active and passive cooling of PVs and electronic devices, etc.[3–5]. Thermal energy can be stored via two basic techniques: Sensible Heat Storage and Latent Heat Storage. In the former technique, the temperature of the storage material varies with stored energy, whereas in the latter, the storage material stores energy via phase change (i.e. steam-water, ice-water, etc.)[6–8].

The high latent heat of the PCMs and their relatively lower phase change temperature are the core properties which have interested researchers for the aforementioned applications[9] . Despite key desirable properties of PCMs (high energy-storage density and isothermal characteristics)[10], their low thermal conductivity[11,12] remains the major bottleneck for many applications. This paper investigates the effect of various carbon-based hybrid nano-fillers with different physical dimensions and features on the thermal conductivity and diffusivity of paraffin-based PCM nanocomposites.

2. Experimental

2.1. Materials

IOP Conf. Series: Materials Science and Engineering **380** (2018) 012008 doi:10.1088/1757-899X/380/1/012008

PCM used for the purpose of all experiments is Paraffin wax purchased from Rubitherm (RT42). Nanofillers used for the composite preparation included Graphene Nanoplatelets (Alfa Aesar 47132), Multi-Walled Carbon Nanotubes (MWCNTs) (Sigma Aldrich 724769), Carbon Nanofibers (CNFs) (Sigma Aldrich 719781), and TRG (Sixth Element SE1432). All materials were utilized as supplied without further treatment or purification.

2.2. Preparation of composite PCM

A pure GNP/PCM composite was first prepared by melt mixing aided with sonication using pure GNP as the nanofiller to observe the resultant thermal properties of the PCM and to serve as a control sample to observe the effect of hybrid-nanofillers. The hybrid composite samples were prepared using an overall carbon content of 10 wt.% while varying the composition of the hybrid filler (i.e. CNTs/CNFs/TRG) from 0-2 wt.%. Table Synthesis route used to prepare the hybrid PCM composites was melt mixing aided with sonication. A known mass of paraffin wax was melted in a vial and placed in a temperature controlled bath sonicator at 55°C. Hybrid filler (0-2 wt.% CNT/CNF/TRG) was added to the vial containing the melted wax and sonicated for 30 minutes, followed by addition of the remaining GNP as the base nano-filler. The melted mix was again sonicated for 30 minutes while the bath sonicator's temperature was maintained well above 50C. The nanocomposite was then quenched in an ice bath and used as is for characterization.

Table 1. Summary of compositions of the hybrid nano-fillers in PCM composites.

Sample Code	PCM (wt.%)	GNP (wt.%)	Hybrid* (wt.%)
H-XXX-01	90	9.75	0.25
H-XXX-02	90	9.50	0.50
H-XXX-03	90	9.00	1.00
H-XXX-04	90	8.00	2.00

Where XXX is the filler code i.e. CNT for Carbon nanotubes, CNF for Carbon Nanofibers and TRG for Thermally Reduced Graphene

2.3. Characterization

2.3.1. Thermal storage properties.

Thermal properties, i.e. melting/solidification temperatures (Tm/Tf), latent heat capacity (ΔHm/ΔHf), and specific heat capacity (Cp), of the PCM composite were analyzed using DSC (Q2000, TA Instruments). DSC runs consisted of an initial melting/solidification cycle to remove the effect of uneven heating on the specimens, the second cycle was then used to record aforementioned thermal data. The DSC analysis was performed using a heating rate of 5°C/min in a temperature bracket of 5 – 80°C. Thermal cycling was also performed to observe the effects of repeated heating/cooling cycles on the PCM stability, a total of 40 cycles were run. TGA (TGA550, TA Instruments) was used to observe the effect of fillers on the stability of PCM. The thermal conductivity of the samples was measured by a Hot Disk thermal analyzer (Hot Disk 2500, Sweden) using the Transient Plane Source (TPS) method based on a transient technique. According to this method[13]a disk-shaped TPS sensor is placed between two circular sample pieces.

2.3.2. Morphology characterization.

Scanning electron microscopy (SEM) imaging was used to observe the morphology of the nano-fillers and qualitatively assess the dispersion and distribution of the nano-fillers within the composite materials. The high intensity of the SEM at higher magnifications resulted in the phase change of samples therefore a lower magnification was used qualitative assessment of the composite structure.

IOP Conf. Series: Materials Science and Engineering **380** (2018) 012008 doi:10.1088/1757-899X/380/1/012008

3. Results and discussion

3.1. Morphology and dispersion of Nano-fillers in composite PCMs

SEM images of the PCM composites revealed the dispersion and distribution of the nano-fillers within the composite. The images of nano-fillers after being dispersed in the PCM were taken with 2.00 wt% samples used as representative samples for a qualitative assessment of each nano-filler batch. SEM images of the GNP-CNF hybrid are presented in Figure 1 which shows the remarkable dispersion of the CNF in the composite with minimal agglomeration. CNT dispersion (see Figure 2) is similar to CNF; however, they appear to be more entangled unlike the CNFs which appear as separate individual strands.

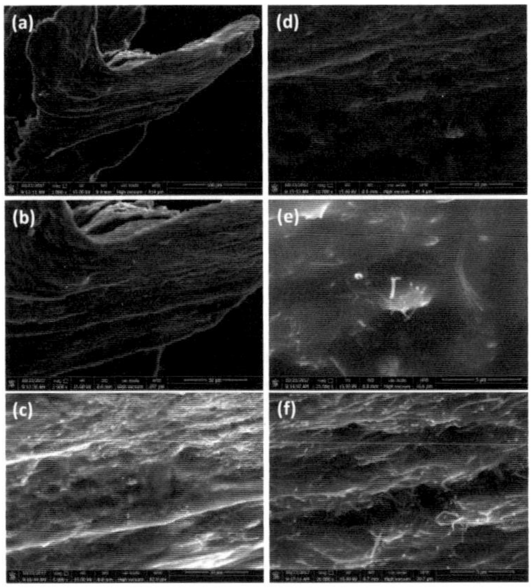

Figure 1. (a)-(f) SEM images of GNP-CNF hybrid nano-filler PCM composite.

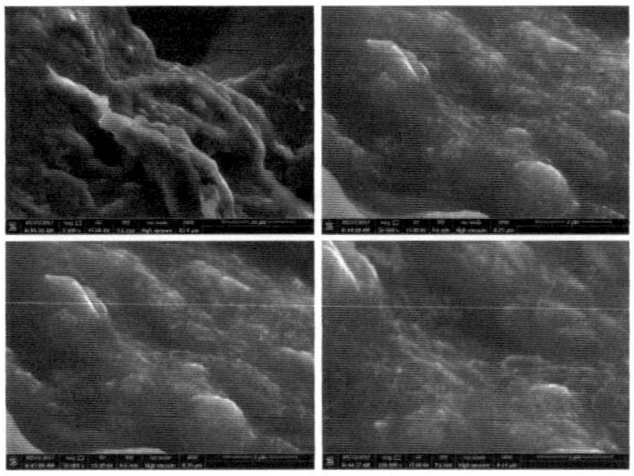

Figure 2. SEM images of GNP-CNT hybrid nano-filler PCM composite.

3.2. Thermal properties of composite PCMs

Thermal properties of the composite PCM were characterized using DSC. The melting/solidification onset temperature (Tm/Tf), latent heat of melting/solidification (ΔHm/ΔHf), and heat capacities for solid and liquid phases were extracted from the DSC heating and cooling curves obtained as per method described in experimental section and are provided in Table 3 (appendix).

The addition of nano-fillers serves an important role in increasing the thermal conductivity; however, it does so at the expense of the energy storage capacity as it does not add to the latent heat of fusion. Moreover, DSC analysis reveals the adverse effect of the addition of these hybrid nano-fillers on the phase change temperatures and enthalpies of the hybrid-nanocomposites. DSC heating/cooling curves of the pure and hybrid composites are presented in Figure 3 and Figure 4 . The two characteristic paraffin peaks (solid-solid and solid-liquid phase change peaks) are visible and the addition of fillers contributes to a lowering of both the peaks. The peak temperatures during melting were around 18°C and 43°C whereas in the cooling cycle the peak temperatures were noted at 17°C and 42°C. As expected, melting temperatures and enthalpies are marginally lowered due to nanofiller addition; however, no obvious relationship can be established between the loading and the temperature. A qualitative assessment of the shrinking and lateral translation of DSC melting/cooling peaks also substantiates this observation. A quantitative comparison can be observed in Figure 5 and Figure 6 for the solid-liquid phase change onset temperatures and enthalpies. It was observed that the decrease in peak solid-liquid enthalpies for CNF, CNT and TRG were 16%, 14% and 12% respectively from a pure wax enthalpy of 135 J/g.

IOP Conf. Series: Materials Science and Engineering **380** (2018) 012008 doi:10.1088/1757-899X/380/1/012008

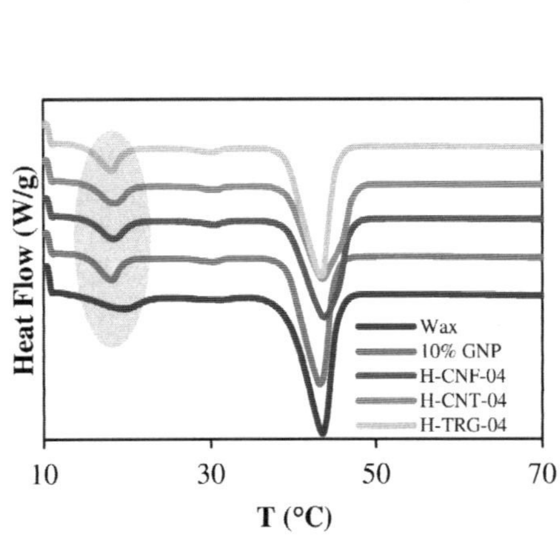

Figure 3. DSC Heating Curves for various filler types at maximum hybrid loading.

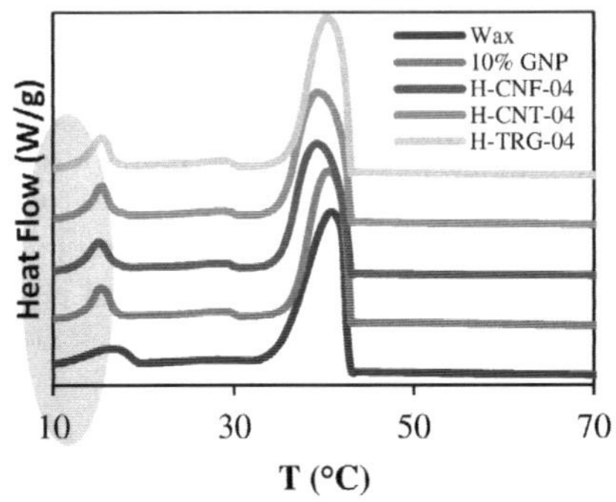

Figure 4. DSC Cooling curves for various filler types at maximum hybrid loading.

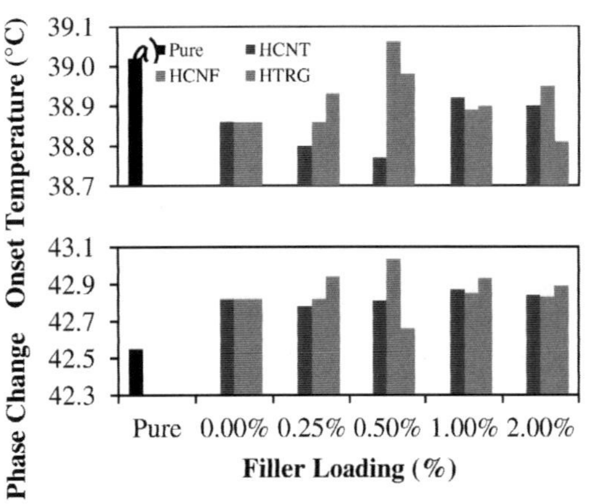

Figure 5. Variation of phase change onset temperature as a function of hybrid filler composition, a) melting and b) solidification process.

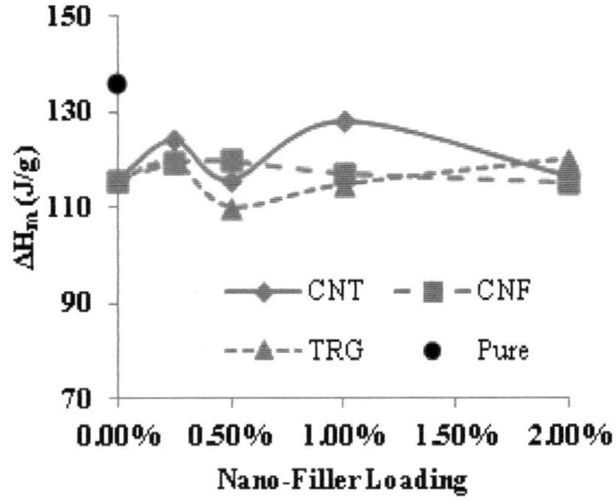

Figure 6. Latent heat of composite PCMs as a function of nano-filler loading during melting cycle.

3.3. Thermal conductivity of PCM

Solid phase thermal conductivity and diffusivity of the hybrid filler nanocomposites was measured as a function of the hybrid filler loading (0 – 2 wt.%). Measurements were performed as per procedure described in the experimental section and two readings were obtained for each sample point later averaged (average difference between values was 1%). Fig.7 summarizes the results, where inset a) presents the actual conductivity and diffusivity data, inset b) and c) presents the thermal conductivity and diffusivity enhancement (%) with the pure wax as the base on the left axis, and the enhancement with 10% GNP as the base on the right axis.

Results revealed that replacement of core GNP filler with CNF, CNT and TRG hybrids resulted in an overall increase in thermal conductivity. However, the enhancement in conductivity is less pronounced for TRG at 2 wt.% loading in comparison to CNT and CNF, whereas CNF dominates in promoting thermal enhancement. An overall increase by 143%, 158% and 174% in thermal conductivity and 179%, 193% and 214% in thermal diffusivity was observed at 2 wt.% hybrid loading of TRG, CNT and CNF respectively. There was an anomaly at 1 wt% loading which had slightly higher thermal conductivity increase for TRG resulting at 158%. This could be attributed to the TRG facilitating the

IOP Conf. Series: Materials Science and Engineering **380** (2018) 012008 doi:10.1088/1757-899X/380/1/012008

formation of a connected network of filler resulting in a more effective thermally conductive network within the composite.

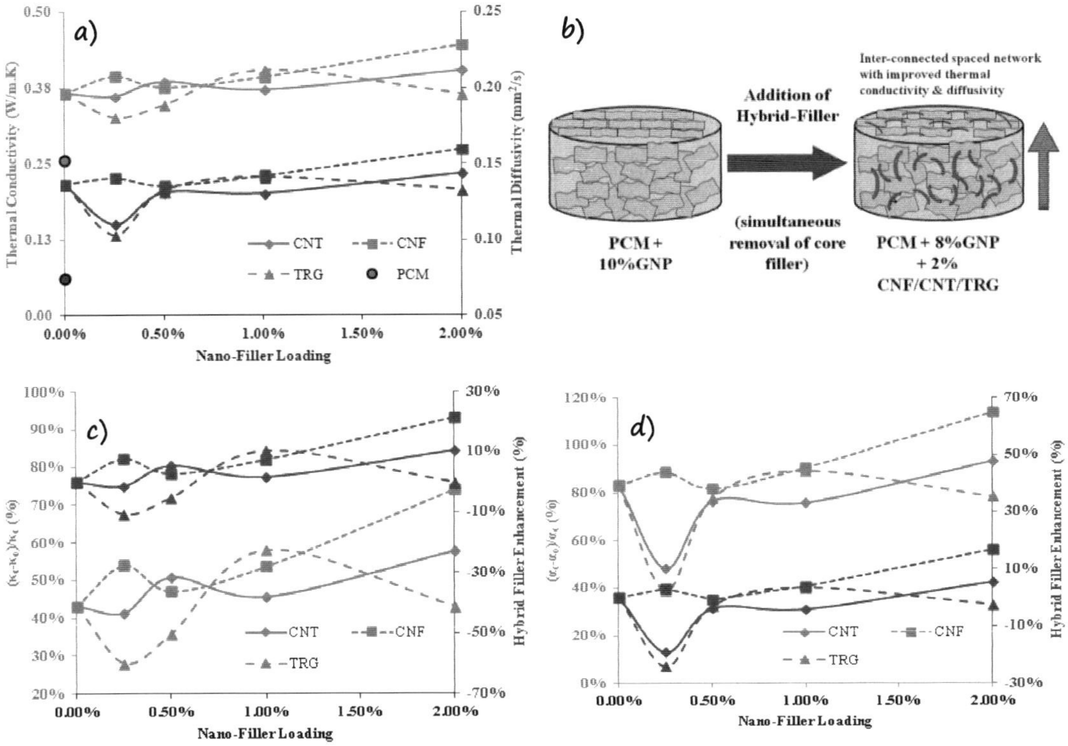

Figure 7. a) Results of measured thermal conductivity for various filler loadings, b) schematic of the proposed effect of the hybrid nanofiller on composite properties, c) & d) percentage enhancement in thermal conductivity ($\boldsymbol{\kappa}$) and diffusivity ($\boldsymbol{\alpha}$) with reference to pure wax (left axis) and 10% GNP composite (right axis).

It is interesting to observe such a remarkable increase in the overall thermal conductivity despite the removal of 2 wt% of GNP which is known to have a planar structure and high contribution to thermal enhancement within composites. It is proposed that the increased thermal conductivity with the introduction of hybrid fillers allows for the spacing of the core GNP filler. This spacing and creation of a more porous and widespread network allows for a more thermally conductive network to be achieved, thus increasing the thermal conductivity and diffusivity (See Fig for schematic of the proposed process). This theory is further substantiated by the fact that though TRGs are inherently less thermally conductive than the GNPs, despite the introduction of 2 wt% TRG and simultaneous removal of the same amount of GNP maintained the net gain in thermal enhancement. CNFs' inherent structure (i.e. dimensions) and intrinsic thermal properties allow it be a prime candidate for hybrid-filler for thermal conductivity enhancement, followed by CNTs.

To further observe the contribution of the hybrid-fillers, the thermal enhancement was compared to a control sample with 10% GNP only. The relative enhancement was evaluated using the formula $(\kappa_c - \kappa_o)/\kappa_o$, where κ_c is the bulk composite thermal conductivity and κ_o the control sample (i.e. 10% GNP/pure PCM). Though at low loadings there was no significant thermal enhancement, however, at 2 wt.% loading CNTs and CNFs showed an increase of 122% and 110% respectively. No improvement was observed for TRG, however, it is still interesting as there was a reduction in the more conductive filler GNP and yet the overall conductivity was the same. It should be noted that the

IOP Conf. Series: Materials Science and Engineering **380** (2018) 012008 doi:10.1088/1757-899X/380/1/012008

introduction of highly-conductive hybrid-fillers is not alone responsible for the improved thermal conductivity, in fact, as demonstrated by Babaei et al.'s[14] molecular dynamic simulation results, the filler-induced alignment of matrix molecules i.e. PCM, is what results in the intrinsic thermal conductivity of the composite.

4. Conclusions

It can be concluded from the results of this work that the hybrid nano-fillers can have a significant contribution to the thermal enhancement of the overall nano-composites. As presented in the results, an overall increase by 143%, 158% and 174% in thermal conductivity and 179%, 193% and 214% in thermal diffusivity was observed at 2 wt.% hybrid loading of TRG, CNT and CNF respectively (pure PCM as ref.). Furthermore, the CNF and CNT hybrid-fillers contributed an increase of 122% and 110% in the thermal conductivity and 117% & 105% in thermal diffusivity of the composite PCM (10%-GNP as ref.). The improved thermal enhancement is attributed to the inter-connected spaced out filler network within the composite matrix. Future work should explore ways to make these composite form-stable without compromising the thermal conductivity of the final composite PCM.

5. Acknowledgement

Authors would like to acknowledge the financial support and research facilities provided by Hamad Bin Khalifa University and Qatar Foundation. Also Dr. Igor and Centre for Advanced Materials, Qatar University for providing raw materials and characterization support for this research.

6. References

[1] Martín-Pomares L, Martínez D, Polo J, Perez-Astudillo D, Bachour D and Sanfilippo A 2017 Analysis of the long-term solar potential for electricity generation in Qatar *Renew. Sustain. Energy Rev.* **73** pp 1231–46

[2] Duan Z, Zhang H, Sun L, Cao Z, Xu F, Zou Y, Chu H, Qiu S, Xiang C and Zhou H 2014 CaCl2·6H2O/Expanded graphite composite as form-stable phase change materials for thermal energy storage *J. Therm. Anal. Calorim.* **115** pp 111–7

[3] Sharma A, Tyagi V V, Chen C R and Buddhi D 2009 Review on thermal energy storage with phase change materials and applications *Renew. Sustain. Energy Rev.* **13** pp 318–45

[4] Shaikh S and Lafdi K 2010 C/C composite, carbon nanotube and paraffin wax hybrid systems for the thermal control of pulsed power in electronics *Carbon N. Y.* **48** pp 813–24

[5] Shahil K M F and Balandin A A 2012 Graphene–Multilayer Graphene Nanocomposites as Highly Efficient Thermal Interface Materials *Nano Lett.* **12** pp 861–7

[6] Goli P, Legedza S, Dhar A, Salgado R, Renteria J and Balandin A A 2014 Graphene-enhanced hybrid phase change materials for thermal management of Li-ion batteries *J. Power Sources* **248** pp 37–43

[7] Hasnain S M 1998 Review on sustainable thermal energy storage technologies, Part I: heat storage materials and techniques *Energy Convers. Manag.* **39** pp 1127–38

[8] Farid M M, Khudhair A M, Razack S A K and Al-Hallaj S 2004 A review on phase change energy storage: materials and applications *Energy Convers. Manag.* **45** pp 1597–615

[9] Mehrali M, Latibari S T, Mehrali M, Indra Mahlia T M and Cornelis Metselaar H S 2013 Preparation and properties of highly conductive palmitic acid/graphene oxide composites as thermal energy storage materials *Energy* **58** pp 628–34

[10] Abhat A 1983 Low temperature latent heat thermal energy storage: Heat storage materials *Sol. Energy* **30** pp 313–32

[11] Karaipekli A, Biçer A, Sarı A and Tyagi V V 2017 Thermal characteristics of expanded perlite/paraffin composite phase change material with enhanced thermal conductivity using carbon nanotubes *Energy Convers. Manag.* **134** pp 373–81

[12] Zhang L, Zhang Y, Xu H, Wang H and Du Q 2016 Polymer/graphene oxide composite microcapsules with greatly improved barrier properties *RSC Adv.* **6** pp 7618–25

IOP Conf. Series: Materials Science and Engineering **380** (2018) 012008 doi:10.1088/1757-899X/380/1/012008

[13] Coquard R, Coment E, Flasquin G and Baillis D 2013 Analysis of the hot-disk technique applied to low-density insulating materials *Int. J. Therm. Sci.* **65** pp 242–53
[14] Babaei H, Keblinski P and Khodadadi J M 2013 Thermal conductivity enhancement of paraffins by increasing the alignment of molecules through adding CNT/graphene *Int. J. Heat Mass Transf.* **58** pp 209–16.

IOP Conf. Series: Materials Science and Engineering **380** (2018) 012009 doi:10.1088/1757-899X/380/1/012009

Low energy solution ball milling of graphene nanoplatelets (GNPs) reinforced aluminium nanocomposites and its mechanical properties

Z Baig [1], O Mamat [1], M Mustapha [1], S Ali [1], and M Yasir[1]

[1]Mechanical Engineering Department, Universiti Teknologi PETRONAS, 32610, Bandar Seri Iskandar, Perak, Malaysia

Email: meteng81@gmail.com, drothman_mamat@utp.edu.my

Abstract. Graphene Nanoplatelets (GNPs) possess outstanding properties which can be utilized to reinforce Al metal for various applications of automotive and aerospace sectors. The major hurdle for real-time applications is the GNPs dispersion in Al matrix which is a challenging task. To address this issue, a combination of processing like solvent dispersion via tip sonication and low energy ball milling at 200 and 300rpm were employed to investigate the GNPs dispersion at various fractions and their effects on final nanocomposite properties. Microstructural analysis of nanocomposite powder was carried out to investigate dispersion analysis. A decrease in relative density with increase in GNPs content showed by all samples. Hardness and wear characterization of the nanocomposite samples were performed. It was found that 0.3wt%GNPs/Al samples has shown maximum increase in hardness (35.61%) and reduce wear rate of (76.68%) than pure Al at 300rpm milling. Finally, GNPs have shown their reinforcing effect by increasing hardness and wear rate by provide effective lubrication.

1. Introduction

Aluminium matrix composites (AMCs) have found interesting applications in the automotive and aerospace sectors where high structural loadings and contact motions are involved [1]. Aluminium matrix composites (AMCs) possess high specific strength, low density, high thermal conductivity and improved tribological properties. Noteworthy, still the AMCs properties are not up to the mark as theoretically calculated due to limitations associated with reinforcements such as their micron size. Therefore, size and type of the reinforcement basically dictate the final properties of the composites. From last few years, aluminium nanocomposite (AMNCs) has attracted a lot of attention due to the superior properties exhibited by them [2]. Graphene, a 21st century wonder material is considered as the ideal reinforcing agent to improve overall properties of the AMNCs to fulfil the increasing remands raised by the potential applications consumers.

Graphene possesses high mechanical properties like high elastic module, high tensile and fracture strength due to its 2D planar sheet morphology and tightly bonded covalently attached carbon in a honeycomb structure and more favourable to load transfer [3]. Importantly, graphene has high surface area and aspect ratio which cause high attractive forces to attract sheets due to van der Waal forces and form aggregates. Due to these reasons, graphene dispersion in metal matrices is a most challenging task. It is well established that uniform dispersion of the graphene in Al matrix ultimately lead to final nanocomposite superior properties and highlighted in many research works [4]. In recent years,

various approaches were adopted to cater the problem graphene agglomeration via novel or modified processing coupling graphene surface modification covalent or non-covalent functionalization [5]. In terms of processing, powder metallurgy (P/M) processing of the graphene/Al nanocomposite under dry conditions emerged as the most successful technique to ease of fabrication, the potential for uniform dispersion and less chemical reactions [6]. The high energy ball milling processing effectively employed to obtain well-dispersed graphene in Al powders but at the expense of graphene structure deterioration due to harsh conditions involved. As a result, defects formed in graphene structure lead to unwanted reactions with metals and drastically effects final properties. Similarly, there are also some other techniques available employed by researchers to control dispersion issue but mostly are time-consuming and have their own limitations such as molecular level mixing and in situ graphene formation on the metal particles [7]. In recent years, a solvent dispersion of the graphene has been adopted combine with other processing techniques [8]. Now solvent dispersion of graphene is considered as the pre-requisite technique for graphene dispersion. Solvent dispersion of the graphene can be achieved by bath or tip sonication which effectively produces quality solvent dispersion because high energy ultrasonic waves effectively break the notorious irresistible attraction of van der Waals forces to restrict agglomerates formation [9].
In this research, an investigation of the graphene dispersion in Al powders and their effect on properties was performed by using combine processing of graphene solvent dispersion by tip sonication and low energy solution ball milling to avoid possible graphene structure impairment. The low energy solution ball milling of the GNPs/Al nanocomposite powders were processed at two rotational speed (200 and 300 rpm) in order to analyse their effect on dispersion and final nanocomposite properties.

2. Materials
Argon gas atomized and spherical shaped pure Aluminium (Al) powder having >99% purity and particle size up to 10-15μm also confirmed by laser diffraction (Mastersizer 2000) as shown in Figure 1a. Al powder was purchased from CNPC Powder group Co. Ltd., China. High quality Graphene nanoplatelets (GNPs) was supplied by U-Gent technology, Malaysia contains average thickness of GNPs was 5-10nm equals to 15 to 20 graphene sheets per stack of GNPs. The size equals to 6μm-26μm with carbon > 99.5 wt % as depicted in Figure 1b showing stacks of graphene layers. GNPs showed density of 2.17 g/cm3 and Al powder have 2.700 g/cm^3 as per measured with helium pycnometer.

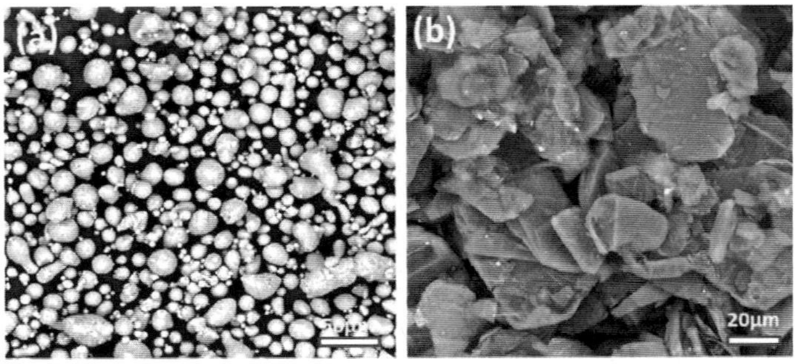

Figure 1. SEM micrographs of (a) as-received Al power and (b) Graphene nanoplatelets (GNPs).

3. Methodology
A facile low energy blending also known as tumbling ball milling was used to incorporate GNPs into the Al powder using solution blending. Firstly, dispersion of GNPs was achieved in Ethanol solvent using tip sonicator purchased from Qsonica (Model Q500), USA. The GNPs solvent dispersion was attained using optimized sonication parameter equals to sonication time of 30 min and at 60%

amplitude [10]. Different GNPs dispersion solutions were prepared using various GNPs factions such as 0.1, 0.3, 0.5 and 1 wt. %. Secondly, after GNPs solvent dispersion, each solution underwent for solution ball milling for 10 min at 250 RPM to achieve GNPs exfoliation. The 1.0-liter Roalox Alumina-Fortified grinding jar with ZrO_2 balls (10mm) using US Stoneware ball milling machine was used. Thirdly, 50g of Al powder was added into each ground GNPs solution jars with a ball-to-powder ratio of 10:1 and further milling was conducted for 2 hours at 200 and 300 RPM respectively. After every 30 min stays for 5 min was given to avoid heating. After dispersion processing completion, GNPs/Al solutions were vacuum filtered and dried in vacuum oven at 90°C for 10 hours to obtain dry GNPs/Al nanocomposite powder for powder dispersion analysis and consolidation. Pure Al and nanocomposite powders were compacted via uniaxial cold compaction at 500Mpa for 1 min hold time. The obtained compact have dimension equals to 30 mm diameters and 4-5mm thickness. The consolidation of the compact samples was done in box furnace under the N_2 atmosphere at 620°C sintering temperature for 2 hours as dwell time. After completion of process, samples were furnace cooled and taken out for testing.

4. Characterization techniques

Sintered density of the pure Al and nanocomposite sintered samples were measure by using Archimedes' Principle. Relative sintered densities (%) were calculated by dividing sintered density (g/cm^3) to theoretical density of the pure Al and nanocomposite powders. Theoretical density of nanocomposite powders can be find by using rule of mixture as per formula (1) given below [11]:

$$\frac{1}{\rho_C} = \frac{W_m}{\rho_m} + \frac{W_f}{\rho_f} \quad \textbf{(1)}$$

Where, ρ_c = composite theoretical density, ρ_m = density of Al (2.7g/cm^3), ρ_f = density of GNPs (2.17 g/cm^3), W_m and W_f are the weight fractions of the Al matrix and GNPs powders used respectively. Micro hardness of the polished sintered samples were measured using Vickers hardness testing at 200gf for 15 minutes holding time. At least five values were calculated and averaged value was recorded. Dispersion analysis and graphene presence in the Al nanocomposite powders were performed via microstructural analysis using scanning electron microscope (SEM) (Phenom- Pro X) equipped with an energy-dispersive X-ray spectrometer (EDS). Wear testing on the polished samples were performed via a pin on disc under dry condition by using Taber Linear Abraser 5750. The process detail and parameters were used same as previously used in our experiment can be found elsewhere [5].

5. Results and discussions

5.1. Dispersion analysis

Figure 2 and 3 represents the SEM images of the as-received Al and GNPs/Al nanocomposite ball milled powders processed at two rotational speed (200 and 300rpm). As can be seen that as-received Al powders have spherical shape morphology (Figure 2a). However, interesting results were showed by the nanocomposite powders at both rotational speeds. Figure 2 shows the various content of the GNPs dispersion in ball milled Al powders. It can be seen that at 200rpm GNPs dispersed only up to 0.1wt% content due to less involvement of the impact forces required for uniform dispersion. While other fractions of the GNPs present as agglomeration in the Al powders as shown in Figure with 1 wt. % GNPs content. It should also be noted that due to low energy rpm, Al powder morphology has remained same i.e. spherical which depicts no effect of balls impact during the milling process. This may be the reason for low content dispersion of GNPs in Al powders.

Similarly, Figure 3 presents dispersion state of the different GNPs content in Al powder after solution ball milling at 300rpm. At this rotational speed, GNPs were uniformly dispersed up to 0.3wt% as compared to other GNPs content as seen in Figure 3. Such behaviour can be related to the increase in

IOP Conf. Series: Materials Science and Engineering **380** (2018) 012009 doi:10.1088/1757-899X/380/1/012009

rotational speed which readily disperses the GNPs into the Al powder up to 0.3wt% and others content showed agglomerations availability in the Al powders. This can also be confirmed by the change of the morphology of the Al powders under 300rpm i.e. transform to flake shape reveals deformed morphology. The morphological transformation is more pronounced up to 0.3wt% GNPs and after that flaky shape of powders were not observed due to the GNPs lubrication effect provide the cushion for powder deformation and contribute to the reagglomeration of the GNPs. Overall, it can be found that major dispersion was provided by the solvent dispersion and further dispersion assistance was more provided by ball milling at 300rpm than 200rpm due to increase in impact forces by balls. Further, at a high content, GNPs re agglomeration was severely formed and clearly seen in the micrographs.

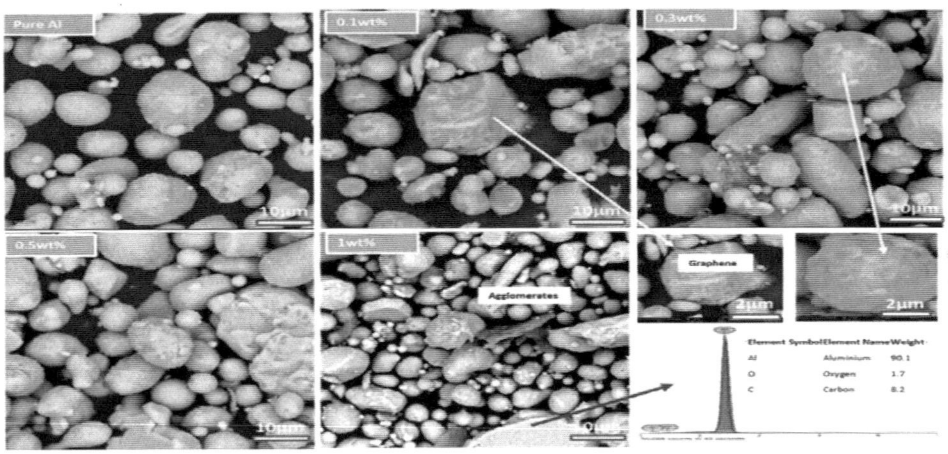

Figure 2. SEM images of the (a) Pure Al and GNPs/Al powder ball milled at 200 rpm at various GNPs fractions (0.1, 0.3, 0.5, and 1 wt. %). The magnified images are also shown along with EDX results showing graphene presence and its agglomerations.

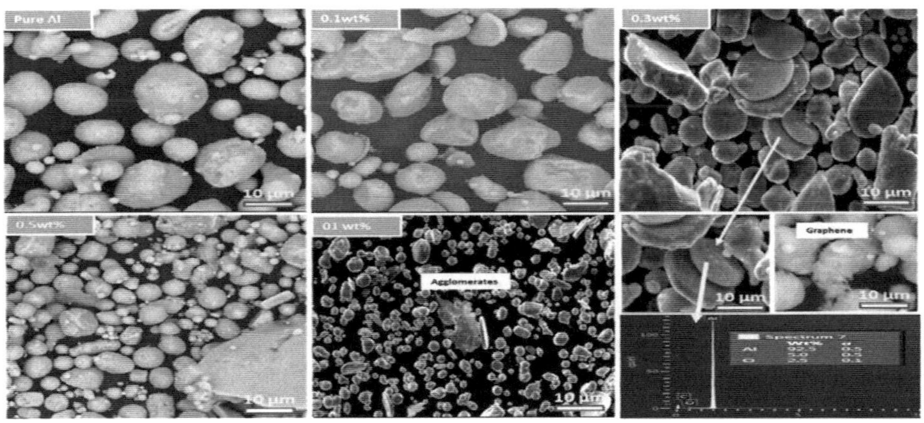

Figure 3. SEM images of the (a) Pure Al and GNPs/Al powder ball milled at 300 rpm at various GNPs fractions (0.1, 0.3, 0.5, and 1 wt. %). The magnified images are also shown along with EDX results showing graphene presence and its agglomerations.

5.2. Density and hardness analysis

Figure 4a illustrates the relative density (%) of the pure Al and sintered GNPs/Al samples at 200 and 300 rpm. As can be seen that at low content both types of GNPs/Al samples density increases as compared to pure Al than abruptly decreases due to the addition of high content of low-density GNPs.

This kind of behaviour also reported by Rashad et al [9]. It can also be related to the formation of agglomeration after GNPs reagglomeration at high content because with increase in content graphene sheets tend to close to each other and attracted to each other thus form agglomerates [12] .Figure 4b represents the hardness behaviour of the GNPs/Al samples processed at two rotational speeds. GNPs/Al samples processed at 200rpm showed the increase in hardness to 45Hv at 0.1wt% GNPs content than pure Al hardness (36Hv) whereas beyond this GNPs content hardness decreases. Similarly, at 300 rpm, GNPs/Al samples exhibit an increase in hardness of 48.82Hv at 0.3wt% GNPs content as compared to pure Al (36Hv) equals to 35.61% increase. The increase in hardness showed by both nanocomposite samples can be attributed to the uniform dispersion of GNPs whereas a decrease in hardness can be ascribed to the agglomeration effect which is in accordance with SEM results (Figure 2 and 3). In addition, GNPs/Al samples processed at 300rpm has shown the highest hardness of about 35.61% and 200 rpm exhibit 25% increase than pure Al. The improvement in the hardness of nanocomposite samples validates the reinforcing effect of GNPs addition to the Al matrix which effectively restricts the dislocations motion and caused grain pinning. Similarly, GNPs/Al samples ball milled at 300rpm has shown the highest increase due to the higher impact of the balls to disperse GNPs at higher content and also due to the morphological change i.e. spherical to flaky shape lead to the higher surface area and results in better sinterability.

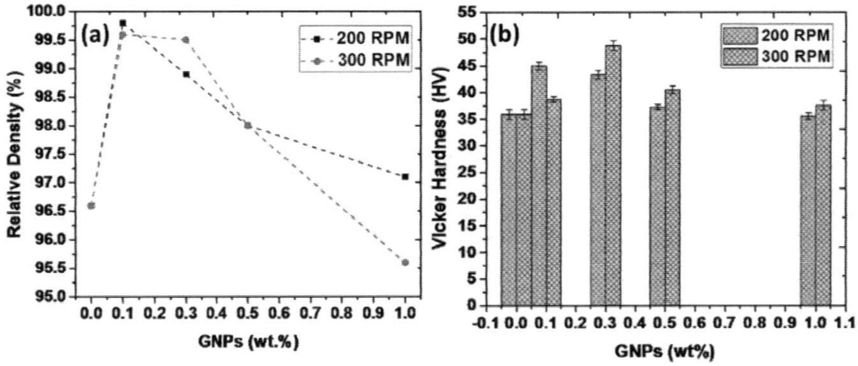

Figure 4. (a) Trends of relative sintered density (%) of Al and GNPs/Al Nanocomposite. (b) Micro hardness response of the pure Al and different fractions of GNPs/Al nanocomposite.

5.3. Wear analysis

Figure 5a and b demonstrate the wear behaviour of the GNPs/Al nanocomposite samples processed via solution low energy ball milling at 200 and 300rpm. Figure 5a reveals the wear loss of GNPs/Al samples at both rotational speeds. It can be seen that samples also exhibit the same as shown by hardness behaviour because hardness and wear have proportional relationship as reported earlier [13]. Therefore, GNPs/Al processed at 200rpm showed lowest wear loss and ultimately wear rate (Figure 5b) at 0.1wt% GNPs fraction than pure Al and after this fraction wear rate increases. Similarly, at 300rpm, GNPs/Al displayed lowest wear loss and wear rate at 0.3wt% GNPs fraction. It should be noted that wear behaviour of all nanocomposite reduced than pure Al. It can be seen that GNPs/Al samples processed at 300 rpm have lowest wear loss and wear rate equals to 76.68% and at 200rpm GNPs/Al showed 80.92% reduction in wear rate than pure Al. The decrease in wear rate of the nanocomposite samples can be endorsed to GNPs lubrication effect and uniform dispersion in Al matrix whereas higher wear rate represents the effect of GNPs agglomeration. During the sliding motion of the mating surfaces, GNPs readily comes up on the surface of nanocomposite after removal of oxide layer and form intact lubricant layer protect the surface from wear. Thus, uniform dispersion of the GNPs in Al matrix ensured improved wear response as can be seen in both composites but at low GNPs content.

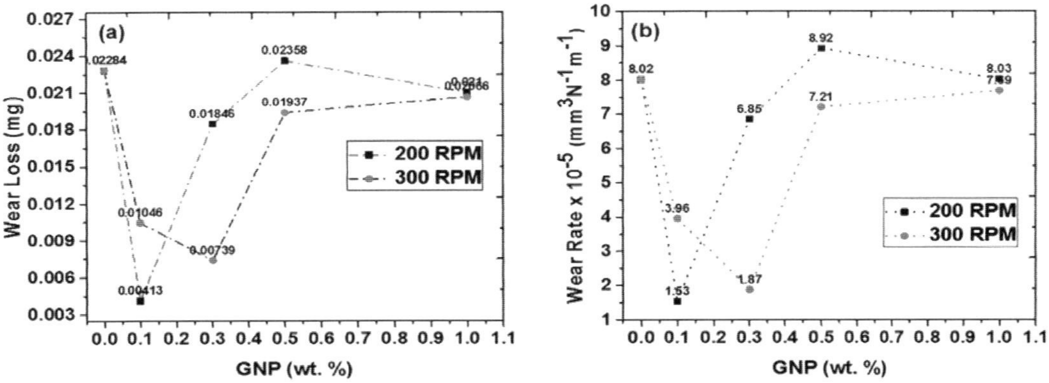

Figure 5. (a) Trends of wear loss (g) of Al and GNPs/Al Nanocomposite. (b) Wear rate response of the pure Al and different fractions of GNPs/Al nanocomposite.

6. Conclusions

In this research, graphene nanoplatelets (GNPs) at various fractions were dispersed in the Al powders via tip sonication solvent dispersion and low energy ball milling performed at two rotational speed (200 and 300 rpm). Consolidation was done by cold compaction and pressureless sintering. Analyzing the SEM results of nanocomposite powders processed at 200 and 300 rpm showed that GNPs were uniformly dispersed at low content i.e. 0.1wt% and 0.3 wt.% respectively. It has been shown that agglomerations still exist at higher content. Relative density (%) of all the composite decreases with increase in GNPs content. GNPs/Al samples processed at 200rpm exhibit 25% at 0.1wt% and at 300rpm display increase of 36% in hardness than pure Al. Similarly, wear rate also followed the same trend at both rotational speeds due to the GNPs lubrication effect and uniform dispersion at low content. Further improvement in the dispersion at higher content will be planned via GNPs surface modification in the Al powder along with compaction and sintering behaviour of the developed Al nanocomposite.

7. References

[1] A. D. Moghadam, E. Omrani, P. L. Menezes, and P. K. Rohatgi, "Mechanical and tribological properties of self-lubricating metal matrix nanocomposites reinforced by carbon nanotubes (CNTs) and graphene–A review," *Composites Part B: Engineering,* **77**, pp. 402-420, 2015.
[2] Z. Hu, G. Tong, D. Lin, C. Chen, H. Guo, J. Xu, *et al.*, "Graphene-reinforced metal matrix nanocomposites–a review," *Materials Science and Technology,* pp. 1-24, 2016.
[3] N. A. Koratkar, *Graphene in Composite Materials: Synthesis, Characterization and Applications*: DEStech Publications, Inc, 2013.
[4] Z. Baig, O. Mamat, and M. Mustapha, "Recent Progress on the Dispersion and the Strengthening Effect of Carbon Nanotubes and Graphene Reinforced Metal Nanocomposites: A Review," *Critical Reviews in Solid State and Materials Sciences,* **41**, pp. 1-46, 2016.
[5] Z. Baig, O. Mamat, M. Mustapha, and M. Sarfraz, "Influence of surfactant type on the dispersion state and properties of graphene nanoplatelets reinforced Aluminium matrix nanocomposites," *Fullerenes, Nanotubes and Carbon Nanostructures,* **25**, pp. 545-557, 2017.
[6] A. Das and S. P. Harimkar, "Effect of Graphene Nanoplate and Silicon Carbide Nanoparticle Reinforcement on Mechanical and Tribological Properties of Spark Plasma Sintered Magnesium Matrix Composites," *Journal of Materials Science & Technology,* **30**, pp. 1059-1070, 2014.
[7] C. Zhao, "Enhanced strength in reduced graphene oxide/nickel composites prepared by molecular-level mixing for structural applications," *Applied Physics A,* **118**, pp. 409-416, 2015.

[8] Z. Baig, O. Mamat, M. Mustapha, A. Mumtaz, M. Sarfraz, and S. Haider, "An Efficient Approach to Address Issues of Graphene Nanoplatelets (GNPs) Incorporation in Aluminium Powders and Their Compaction Behaviour," *Metals,* **8**, p. 90, 2018.

[9] D. W. Johnson, B. P. Dobson, and K. S. Coleman, "A manufacturing perspective on graphene dispersions," *Current Opinion in Colloid & Interface Science,* **20**, pp. 367-382, 2015.

[10] Z. Baig, O. Mamat, M. Mustapha, A. Mumtaz, K. S. Munir, and M. Sarfraz, "Investigation of tip sonication effects on structural quality of graphene nanoplatelets (GNPs) for superior solvent dispersion," *Ultrasonics Sonochemistry,* **45**, pp. 133-149, 2018.

[11] R. M. German and S. J. Park, *Handbook of mathematical relations in particulate materials processing: ceramics, powder metals, cermets, carbides, hard materials, and minerals* **3**: John Wiley & Sons, 2009.

[12] K. S. Munir, Y. Li, D. Liang, M. Qian, W. Xu, and C. Wen, "Effect of dispersion method on the deterioration, interfacial interactions and re-agglomeration of carbon nanotubes in titanium metal matrix composites," *Materials & Design,* **88**, pp. 138-148, 2015.

[13] H. Prashantha Kumar and M. Anthony Xavior, "Effect of graphene addition and tribological performance of Al 6061/graphene flake composite," *Tribology-Materials, Surfaces & Interfaces,* pp. 1-10, 2017.

Preparation and characterization natural rubber-urea-tablets coated by chitosan

M Adlim[1], F Zarlaida, R F I Rahmayani, and R Wardani

[1]Chemistry Department, FKIP, Syiah Kuala University, Darussalam Banda Aceh, Indonesia

Abstract. Excessive urea leaching causes water pollution and fabricating slow release coated urea is usually performed at high temperature or using synthetic polymers and toxic solvent which might generate environmental pollution. This study aims to formulate "green process" of urea slow release tablets and characterize urea slow release tablets. The formula was composed of rice husk ash, urea, and liquid natural rubber (LNR). Several proportions were studied and characterized based on the stability in water and urea leaching. The immobilizing urea was carried in two models; core-shell and mixed model composite either coating or uncoated with chitosan. The slow release urea tablets were stable in water without decomposition at least 9 days of incubation times depending on LNR content but at least 15 mL of LNR was required to obtain stable tablets. Urea leaching from core-shell tablet was 0.045% and 0.75% for mixed model tablet in 10 days incubation times. Chitosan coating did not affect significantly on urea leaching for core-shell model but it caused urea release faster in mixed model after 12 days incubation times. Low LNR and chitosan coating retarded fungus growth on the urea slow release tablet.

1. Introduction

Plants absorb nutrients from the soil especially nitrate in form of water soluble ions. Nitrate as the source of nitrogen is often insufficient naturally for plant growth and urea fertilizer was used as nitrogen supply since it has high nitrogen content (~ 46 wt% in the molecules), affordable, widely available and compatible in uses. The urea was decomposed into ammonium ions and subsequently converted to nitrate catalysed by nitrosomonas and nitrobacter [1-3]. However, urea has low effectiveness, such as hygroscopic, easily denitrified, volatile, rapidly decomposing, leaching, and if accumulated in the ecosystem will adversely affect environmental health [4-5]. Several studies have shown that 40-70% of nitrogen in fertilizers released into the environment without being absorbed by plant roots [6-7].

One of promising alternatives to increasing fertilizer efficiency, reducing the soil toxicity, and minimizing potential negative effects such as overdose is using a slow release fertilizer (SRF) [8]. This type of urea fertilizer is usually fabricated by coating the composite SRF with a synthetic polymer. SRF can reduce the rate of fertilizer release in soil or in water. However, some slow release fertilizers were reported to have weaknesses such as complicated manufacturing, expensive production costs, non-biodegradable polymer matrices, and other less environmentally friendly mixtures [9].

Several studies on preparation slow release urea have been reviewed by several authors [3, 9]. Urea slow release was prepared either by polymer coating using a resin or thermoplastic materials or by inorganic coating such as melt sulphur. Several other coating alternatives are being intensively studied. Coating techniques was reported by using 26% gypsum–sulphur (20% total coating) or 3% paraffin wax [7]. Many other methods have been also introduced including blending urea with urea-formaldehyde polymer [10] and by blending with leftover rice-urea-polyacrylic acid [11]. However the fabrication using friendly material such as liquid natural rubber (LNR) was still very few in publication. Therefore this study aims to formulate the preparation of slow release urea fertilizer using LNR and rice husk ash (RHA) and the study of the characteristics of urea release and its stability in water.

2. Methods

2.1 Composition screening of slow release urea-Mg (SR-Urea-Mg) tablets

Liquid natural rubber latex (LNR) from *Havea bracileansis* (rubber plant grown in Aceh, Indonesia) was prepared by mixing 2 litre of field rubber latex with 50 mL of ammonium hydroxide solution (2.5%) as soon as after sapping, mixed homogeneously and kept it in a refrigerator before used [12]. Rice husk ash (RHA, 1 kg) was sun-dried, grinned and separated from larger particles by sieving with a 200 mesh sifter. Urea-Mg tablets were prepared from several composition of LNR, water, RHA, Urea and $MgSO_4.7H_2O$. Two tablet models were prepared that were a core-shell model (urea was inside of tablet) and a homogenous mixture model.

The various compositions of LNR in tablet matrix were tested until we found the stable tablet in water. Finally the composition was kept constant where the RHA weight was kept at 60 gram, urea and $MgSO_4.7H_2O$ were 20.8 g and 9.6 g respectively. The composition of LNR was reduced gradually and the water volume was increased accordingly. The mixture was stirred with a mixer to obtain relatively homogenously matrix. The mixture was inserted into a table dies and pressed with a hydraulic press at 100 Nm^2. Then, the urea-Mg tablets were dried at 105°C for 1 hour.

2.2 Decomposition test

At every combination of composite, 10 tablets of SR-Urea-Mg were randomly taken and placed in a 250 ml round bottom boiling flask. It was rotated using a rotary evaporator at 40 rpm for 30 minutes. The higher ratio between stable tablets and the decomposed ones was noted as the best composition formula. Similar to this test, at every combination of components, randomly take 3 tablets were soaked in 100 mL of water. Observation was conducted every day to record the number of tablets decomposed at certain time period. The tablets that lasted without decomposition more than 7 days were considered stable at given composition formula. Having obtained the rational formula, the SR-Urea-tablets was reproduced and coating with chitosan (type ~400000, Sigma) solution (0.6 gram of chitosan in 100 mL of 1.5% acetic acid). The SR-Urea tablet was compared with control that was neither urea, none Mg and without chitosan coating.

2.3 Determination of urea release

One tablet of chitosan-coated SR-urea was randomly taken from 100 dried tablets and soaked in 100 mL of distilled water. This procedure was done three replications and each solution was placed in a water bath to maintain the temperature at around 29°C and covered up with plastic film to avoid evaporation. The solution was stirred gently using a stirring rod and every 3 days, 5 mL of the aliquot was taken for urea determination using a spectrophotometric method. In a separate glass beaker; 0.2015 g of p-dimethylamino-benzaldehyde was dissolved in 0.25 mL of H_2SO_4 (36 N) and 70 mL of CH_3OH (labeled as DB solution). Three milliliters of aliquot were mixed with 2 mL of DB solution and the absorbance was recorded by using the spectrophotometer [13].

Urea standard solutions were made from urea solution at concentration of 1, 2, 3, 4, and 5 ppm. It was taken 3 mL aliquot from each solution was added 2 ml of DB solution, mixed truly and recorded the absorbance at 420 nm using a UV-Vis spectrophotometer model Spekol 2000. The absorbencies were plotted with the concentration to form a calibration curve.

3. Result and discussion

3.1 Composition screening of slow release urea-Mg (SR-Urea-Mg) tablets

Table 1 shows that stability of SR-Urea-Mg tablet increase with increasing of LNR content but it takes higher cost in production since LNR is relatively expensive. Increasing in LNR content let to better solidifying on the tablet matrix and protected urea from leaching into water. Urea tablet containing low LNR caused low adhesiveness toward RHA and not easily to shape the tablet. So the minimum amount of LNR required to obtain stable tablets for 9 days in water incubation was 15 mL. With this proportion, it caused the urea tablet slightly low durability in friction test. About 10% of the dried tablets mechanically decomposed during friction test that mimicking the tablet endurance level during packing and transportation (Table 2).

3.2 Urea release

In the first experiment, urea was placed inside the tablet and the tablets were either coated with or without chitosan. This formula was noted as a core-shell model tablet. It is found that there was no significant different in amount of urea release between coated and uncoated tablet with chitosan. The water content of coated tablet was 7.14% and the uncoated one was 3.76%. The total urea content inside the tablet was 0.633 g per tablet. After 10 days of incubation in water, the urea percentage of urea release was very low that was 0.044-0,045% as shown in Figure 1. The release retardation has correlation with tablet composite was made of LNR and RHA. When LNR-RHA became rigid, it inhibited water to enter the tablet and it made urea remained in solid form until water penetrated the composite. RHA inside the tablet also absorbed urea, it prevented urea to leach out from the tablet.

The core model tablet preparation is still considered complicated because urea was inserted manually inside the tablet and an automatic insertion is being developed. Urea release from the tablet model was also too slow since only 0.0027 g urea per day in 0.04 m^2 or 6.75 x 10^{-2} $g.m^{-2}.day^{-1}$ in water. As comparison, urea uptake rate by potato plant was 3.7 x 10^{-2} $g.m^{-2}.day^{-1}$ which is very close to urea release core-shell model tablet [14]. The polymer coated urea (PCU) and the epoxy resin coating of sulphur coated urea (PSCU) has N release rate of 22.4% and 17.2% respectively in 30 days during cotton planting [15].

Table 1. Comparison of the composit formula and the stability (RHA = 60 g; Urea = 20 g; Mg = 9.6 g)

No	LNR (mL)	Water (mL)	Texture	Dried test (# stable tablet, %)	Wet test (# day standing)
1.	60	0	easy to shape	100	> 9
2.	40	20	easy to shape	100	> 9
3.	20	40	easy to shape	100	> 9
4.	15	45	easy to shape	90	9
5.	10	50	not easy to shape	10	6

The second formula, urea was mixed truly tablet with RHA and LNR in tablet formation. This is named as mixed model tablet. The LNR proportion was gradually reduced up to 15 mL in 60 gram of RHA as shown in Table 1 & 2. The amount of urea per tablet was 0.625 g. The mean water content of coated urea tablet was 36%.

It can be deduced from Figure 2 that percentage of urea release from the mixed model tablet was much higher than from the core-shell model tablet (Figure 1). There was not much different on urea release between coated and uncoated tablet except after 12 days of water incubation times as shown in Figure 2. Obviously, urea release was higher in coated tablet than the uncoated one. This is different from hypothesis that coating was expected to reduce urea release from the tablet. The explanation might be referred from chitosan properties that it absorbed urea and accumulated into the surface then subsequently brought urea to leach out into water.

Table 2. Data on durability test

No	LNR content (mL) in composite	Physically appearance during shaking times	
		First 30 minutes	Last 30 minutes
1	60	Intact	Intact
2	40	Intact	Intact
3	20	Intact	partially crushed
4	15	Intact	partially crushed
5	10	Shattered	shattered

The amount of urea release from core-shell and mixed model tablet was 3.5% (or 1.61% N) in 12 days for chitosan coated tablet and only 0.25% (or 0.11% N) in 12 days for uncoated tablet. This is considered slow since the standardize slow release urea was less than 15% in 24 h or < 75% release in 28 days [9]. Therefore this composite was categorized as slow release urea.

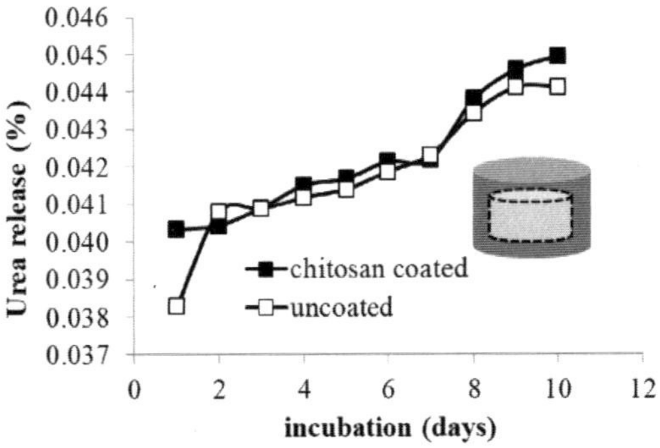

Figure 1. Urea release core-shell model tablet, the comparison of urea release between coated tablet and uncoated tablet during incubation time.

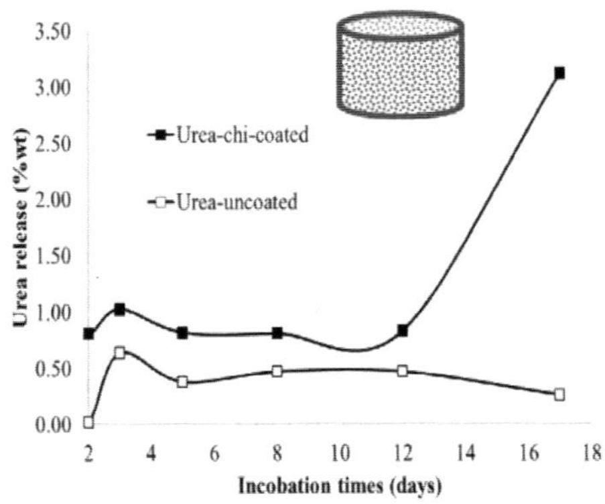

Figure 2. Urea release from mixture model tablet, comparison coated and uncoated chitosan

Chitosan coating prevent the fungi growth in tablet as describe Table 3. Tablet without chitosan coating, cause the tablet surface was covered by fungus hive and less fungus was observed in chitosan coated tablet. This is due to chitosan coating that has antimicrobial properties [16]. The amount of LNR in composition affected the stability and the decomposition times. More stable tablet in water can be achieved by increasing LNR proportion in the formula but it generated more fungi and finally shortened the long life storage. Dried tablet with water content less than 10% extended the long life but drying too long, it caused urea evaporation.

Table 3. The fungi growth observation and decomposition time

LNR content	Chitosan Coating	Fungus growth (**F**) observed & physically decomposed (**D**) (in weeks)																	
		1	2	3	4	5	6	7	8	9	10	11	12	13	14	15	16	17	18
60 g	Yes	-	-	F	D														
(50%)	No	-	F	F	D														
15 g	Yes	-	-	-	-	-	-	-	-	-	-	-	-	-	-	-	-	D	D
(12.5%)	No	-	-	-	-	-	-	-	-	-	F	F	F	D	D	D	D	D	D

With amount of mean water content of 36%, tablet containing high LNR (50%) would have lasted in two weeks before attacked by fungus and subsequently decomposed into powder. Less LNR in composite caused survival longer up to 12 weeks before decompose and it would last up to 16 weeks if it is coated by chitosan film.

4. Conclusion

Two models of stable SL-urea tablets were fabricated from LNR, RHA and chitosan in which minimum 15 mL LNR was required for 60 gram RHA. The more LNR content in the tablet matrix gave the more stable composite but hosted more fungi. With the optimum LNR, the SL-Urea tablet was survive in water for 9 days without decomposing the matrix and the composite also lasted in friction test. Core-shell model SL-urea tablets gave very slow urea release, the rate was slower than the rate nitrate uptake by several plants. Mixed SL-urea tablets performed moderate urea leaching after 12 days in water incubation. Urea release was 0.992%wt per day which is much slower than the existing commercial slow release coated urea (SCU). Low LNR and water content and coating the tablet with chitosan retarded the growth of fungi on the SL-urea tablets. Nutrient releases from these SL-urea tablets into soil shall be studied prior to the fertilizing test performance on the plant. Production cost was also need to be considered.

5. Acknowledgment

We highly appreciated research funding provided by Syiah Kuala University, Aceh Province Indonesia in 2017 financial year with Research Grant number : Nomor : 1442/UN11/SP/PNBP/2017. We also thanks to Baristan (Badan Riset dan Standarisasi) Aceh and Balai Penelitian Ikan Air Payau (BPIP) Ujong Batee, Aceh Province. Indonesia.

6. References

[1] Ghaly AE and Ramakrishnan VV 2013 *Am. J. Agri. Biol. Sci.* **8** 330.
[2] Lubkowski K 2014 *Environ. Eng. Manage. J.* **13** 2573.
[3] Azeem B, KuShaari K, Man ZB, Basit A, and Thanh TH 2014 *Control Release* **181** 11.
[4] Rai N, Ashiya P and Rathore, DS 2014 *Inter. J Innov. Res. Sci. Eng. Technol.* **3** 12991.
[5] Guo M, Liu M, Liang R and Aizhen N 2006 *Journal of Applied Polymer Science* **99** 3230
[6] Hekmat AA, Barati B, Zendehdel M, Norouzi HR and Afraz A 2008 *12th National Chem. Eng. Congress* (Iran: Tbriz).
[7] Ibrahim KRM, Babadi FE and Yunus R 2014 *Particuology* **17** 165.
[8] Xiang Y, Ji-Yun J, Ping H and Ming-Zao, L 2008 *Agriculture Saince China* **7** 469.
[9] Naz MY and Sulaiman SA 2016 *J. Control. Release* **225** 109.
[10] Yamamoto CF, Pereira EI, Mattoso LHC, Matsunaka T and Ribeiro C (2016). *Chem. Eng. J.* **287** 390
[11] Zhou T, Wang Y, Huang S and Zhao Y 2018 Sci. *Total Environ.* **615** 422.
[12] Santipanusopon S and Riyajan S -A 2009 *Physics Procedia* **2** 127.
[13] Husain I, Mahmood Z, Yasmin, R, Jahangir M, Hammed R and Nasir R 2002 *J. Chem. Soc. Pakistan*, **24**, 122.
[14] Zotarellia L, Rens LR, Cantliffe DJ, Stoffella PJ, Gergela D and Burhans D 2015 *Field Crops Res.* **183** 246.
[15] Tian X, Li C, Zhang M, Li T, Lu Y and Liu L 2018 *Soil Tillage Res.* **175** 158.
[16] Goy RC, de Britto D and Assis OBG 2009 *Polímeros: Ciência e Tecnologia* **19** 241.

IOP Conf. Series: Materials Science and Engineering **380** (2018) 012011 doi:10.1088/1757-899X/380/1/012011

Effect of Inhibitors on the Crystallization and Magnetic Properties of Nanocrystalline Alloys

V S Tsepelev[1], **Yu N Starodubtsev**[1,2] **and N P Tsepeleva**[1]

[1]Boris Yeltzin Ural Federal University, 19 Mira str., Yekaterinburg, 620002, Russia
[2]Gammamet Research and Production Enterprise, 92 Tatishchev str., Yekaterinburg, 620028, Russia

E-mail: v.s.tsepelev@urfu.ru

Abstract. In the paper, the crystallization process of nanocrystaline alloys $Fe_{73.5}Cu_1M_3Si_{13.5}B_9$ where M = Nb, W, Mo, V, Cr and the physical origin of different efficiency of impact that inhibitors exert on the structure and magnetic properties has been investigated. It is shown that the structure formed at the peak temperature of the crystallization process is close to the optimal one and ensures high values of magnetic permeability. The efficiency is directly related to the solubility of inhibitory elements in αFe, which in turn affects the diffusivity of atoms. Upon slow migration of grain boundaries, the inhibitory atoms with lower diffusivity, being concentrated near the front of moving boundary, provide stronger drag.

1. Introduction

Nowadays, nanocrystalline soft magnetic materials are widely used in the manufacture of magnetic systems of transformers, electric reactors, chokes, and other devices operated at frequencies up to 10 MHz [1]. High magnetic permeability and low magnetic losses are provided by the nanosized crystal structure of the materials [2, 3]. In the material of the Finemet type with the chemical composition of $Fe_{73.5}Cu_1Nb_3Si_{13.5}B_9$, the average size of Fe-Si crystallites is of the order of 10 nm [4]. The chemical composition of the alloy includes elements that provide realization of an amorphous structure upon quenching (Si, B), as well as elements favoring the formation of nanosized grains in the course of heat treatment (Cu, Nb) [5, 6]. In this alloy, Nb is used for inhibition of grain growth. Other elements can also find their application as inhibitors [7–10], and they can be arranged in the order of increasing the efficiency of grain refinement: Cr, V, W = Mo, Nb = Ta, Zr.

Nanocrystalline soft magnetic alloys are the subject-matter of many physical researches. Of special interest is the process of crystallization from the amorphous solid state and its impact on the magnetic properties of the material, as well as the means of controlling this process. To investigate the process of crystallization, different methods are employed. One of such methods is thermomagnetic analysis. Within the classical thermomagnetic analysis, the temperature dependence of the saturation magnetization is determined [11, 12]. In the alloys with high magnetic permeability, new results can be obtained by analyzing changes in the initial permeability in the course of heat treatment.

In this paper, the thermomagnetic analysis of the nanocrystalline soft magnetic alloys was performed. The physical origin of different efficiency of impact that inhibitors exert on the structure and magnetic properties of nanocrystaline alloys $Fe_{73.5}Cu_1M_3Si_{13.5}B_9$ where M = Nb, W, Mo, V, Cr has been studied.

2. Experimental

The $Fe_{73.5}Cu_1M_3Si_{13.5}B_9$ alloys where M = Nb, W, Mo, V, Cr were melt in a vacuum induction furnace. A 25 μm thick and 10 mm wide ribbon with the amorphous structure was produced by the planar flow casting process. The ribbon was wound up onto ring-type cores with the 32 mm outer diameter and 20 mm inner diameter. The thermomagnetic analysis was performed with the simultaneous recording of the temperature inside the core by a thermocouple and the inductance of the winding wound over the core. For the analysis, 7 cores of $Fe_{72.5}Cu_1Nb_2Mo_{1.5}Si_{14}B_9$ were wound with one and the same ribbon sequentially, one after another. In the course of heat treatment at the temperature of 813 K the cores were taken out of the furnace and quickly cooled at the rate of 70 K/min to fix the structure at different stages of the crystallization, the temperature and inductivity being continuously recorded at the specified time moments shown in Figure 1 by digits 1 – 7. The initial permeability μ was calculated based on the inductance measurements at 1 kHz. Parameters of the DC magnetization curve and hysteresis loop were determined by means of point-by-point DC testing [13]. Specific magnetic losses $P_{0.2/20}$ were measured at the frequency of 20 kHz with the magnetic induction $B_m = 0.2$ T.

3. Results and discussion

Figure 1 shows the in-time changes in the temperature and permeability of a core made of the alloy $Fe_{72.5}Cu_1Nb_2Mo_{1.5}Si_{14}B_9$ upon crystallization in the furnace at the temperature of 813 K. The temperature curve enables determining the initial rate of heating the core prior crystallization, $V_0 = 5$ K/min (a linear portion up to point 1), the onset temperature of crystallization $T_{cr} = 793$ K at point 1 (the starting point of the nonlinear temperature growth), the point of maximal temperature increment – 2, the maximal rate V_m 39 K/min of heating the core after point 2, and the peak temperature in the process of crystallization $T_m = 854$ K at point 3. On the magnetic permeability curve, it is possible to pick out point 4 that corresponds to the beginning of the permeability growth, the portion of the maximal growth rate to point 5, the onset temperature of the maximal permeability growth $T_\mu = 826$ K, the peak value of permeability $\mu_m = 4800$ at point 5, the point of maximal decrease 6, and the stabilized value $\mu_{st} = 840$ at point 7 after completing the crystallization process.

From Figure 1, it is seen that the growth of permeability in the course of crystallization is non-monotonous. After reaching the peak value μ_m, the permeability decrease in time is almost just as quick, going to the steady value μ_{st}. Besides, it follows from Figure 1 that the permeability peak μ_m is 4 minutes behind the temperature peak (T_m).

Table 1 contains data on the permeability $\mu_{0.08}$, the maximum permeability μ_{max}, coercivity H_c, and magnetic losses $P_{0.2/20}$ for the cores the heat treatment of which was interrupted at characteristic time instants depicted by figures 1 – 7 in Figure 1. It follows from Table 1 that of the whole series only cores 1 and 2 are out of line; they correspond to the initial (1) and advanced (2) stages of crystallization, at which the magnetic properties are rather low. Starting from core 3 that corresponds to the peak temperature T_m and further on, for all the subsequent stages, the magnetic properties of the cores are virtually the same. The results of magnetic measurements indicate that after reaching the peak temperature T_m, the alloy structure is nearly optimal.

The bulk of sample 1 is occupied mainly by the amorphous phase in which a crystal phase starts forming. In the micro-diffraction pattern, broad smeared rings are seen and, at their background, the emerging point reflections of the crystal phase. The average size of the grains formed makes up 6 nm. The maximum of the size distribution corresponds to 2 nm; these grains can be ascribed to the crystal phase nuclei. There is also another maximum at about 8 nm in the histogram. Samples 3 and 7 are virtually undistinguished in structure. The micro-diffraction pattern shows that the process of crystallization is completed with the formation of solid solution of silicon in α-Fe. In the electron diffraction patterns, a small amount of the Fe_3Si phase is detected. The histograms of the size distribution show up that the number of the crystallization nuclei in samples 3 and 7 is noticeably decreased compared to sample 1. The main maximum is located at about the average size which is 10 nm, and the second maximum appears to be close to 15 nm. Applying Scherrer's method, the maximum in the vicinity of 10 nm is ascribed to crystals of the α-Fe(Si) solid solution and that in the vicinity of 15 nm, to the Fe_3Si phase.

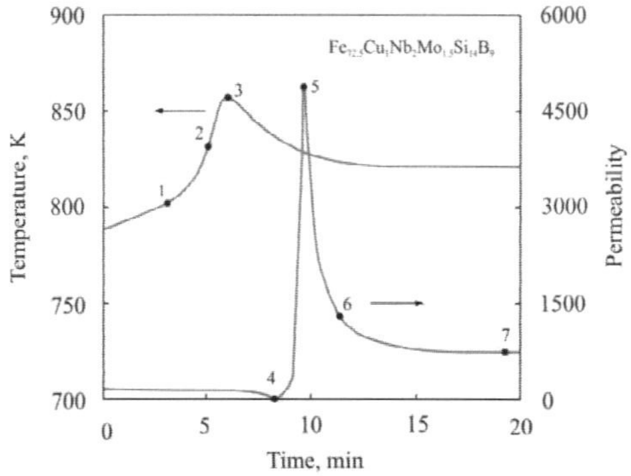

Figure 1. In-time changes of temperature and permeability of the core made from the $Fe_{72.5}Cu_1Nb_2Mo_{1.5}Si_{14}B_9$ alloy in the course of crystallization.

Table 1. Magnetic properties of cores made of the $Fe_{72.5}Cu_1Nb_2Mo_{1.5}Si_{14}B_9$ alloy at 300 K after heat treatment interrupted at different stages of the crystallization process.

Core number	Permeability, $\mu_{0.08}$	Maximal permeability, μ_{max}	Coercivity, H_c, A/m	Magnetic losses $P_{0.2/20}$, W/kg
1	5600	278000	1.25	15.0
2	11700	342000	0.85	8.7
3	67600	521000	0.50	3.3
4	79100	527000	0.55	2.8
5	67000	678000	0.45	4.2
6	73900	575000	0.50	3.2
7	71900	612000	0.50	3.5

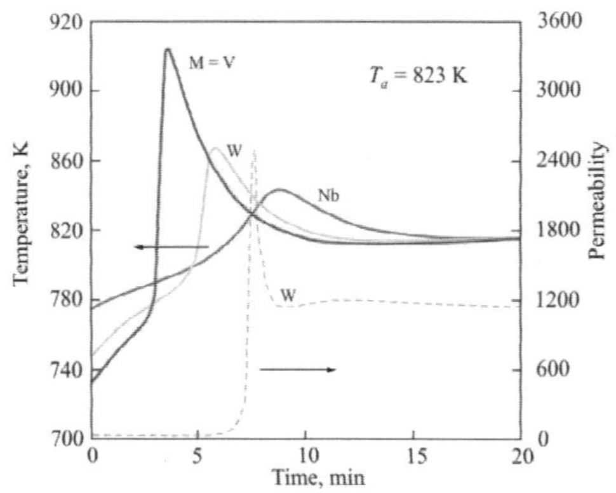

Figure 2. In-time changes of temperature and permeability in the course of crystallization of the core made from the $Fe_{73.5}Cu_1M_3Si_{13.5}B_9$ alloy, where M = Nb, W, V.

Figure 2 displays the in-time change of temperature of cores made from $Fe_{73.5}Cu_1M_3Si_{13.5}B_9$, where M = Nb, W, V in the course of heating in furnace, the temperature of which was kept T_a = 823 K. The temperature curves for different alloys are presented on the same time scale in the range of crystallization. As an example, the dashed line in Figure 2 shows a time dependence of permeability matching the corresponding temperature curve for the $Fe_{73.5}Cu_1W_3Si_{13.5}B_9$ alloy. From the temperature curves there were determined the Curie temperatures of the alloys in the amorphous state T_C, the onset temperature of crystallization T_x, the maximal temperature of the core heating in the course of crystallization T_{max}, and the maximal rate of elevating temperature in the course of crystallization V_{max}. The temperature at which permeability starts growing in the course of crystallization T_μ was determined from the corresponding permeability changes. This temperature can be taken as the Curie temperature of the arising crystalline phase with the composition corresponding to the initial stage of crystallization. All numerical values of the above-listed parameters for different alloys are shown in Table 2. Additionally, the table contains data on the core permeability after annealing, i.e., after crystallization and cooling to 300 K.

Table 2. Parameters of crystallization for the $Fe_{73.5}Cu_1M_3Si_{13.5}B_9$ alloys with the grain growth inhibitors M = Nb, W, Mo, V, Cr.

$Fe_{73.5}Cu_1M_3Si_{13.5}B_9$ alloys with grain growth inhibitors M	Temperature, K				Temperature growth rate V_{max}, K/min	Permeability μ after annealing at 823 K
	T_C	T_x	T_{max}	T_μ		
Nb	600	799	843	822	15	85000
W	597	788	867	847	103	65000
Mo	594	783	868	862	157	75000
V	623	769	922	–	306	1000
Cr	621	763	921	–	445	150

As follows from the Table 2, the maximal permeability is gained on the alloy containing Nb. The W and Mo inhibitors are also efficient for achievement of high permeability, unlike V and Cr. The course of crystallization in the alloys with various inhibitors drastically differs. In the alloy with Nb, the temperature peak is smoothened to a maximal degree, and the process of crystallization is featured by the minimal values of the peak temperature T_{max} = 843 K and the rate of elevating temperature V_{max} = 15 K/min and maximal value of the onset temperature of crystallization T_x = 799 K. The numerical values of T_{max} and V_{max} gradually grow, whereas T_x lowers from row to row sequentially for Nb, W, Mo, V, Cr. The average grain size sequentially grows in the series of inhibitors Nb, W, Mo, V, Cr. In the alloy with Nb the average grain size is 8 mm and no grains larger than 30 nm were detected. In the alloys with V and Cr, the average grain size exceeds 20 nm, the coarsest grains being up to 80 nm.
To discuss the whole bulk of the results obtained, let us present the data on inhibitory elements taken from different sources [14–17]. In Table 3 there are collected solubility of elements in αFe at 1000 K; metallic radius of atoms r_i, relative radius according to the criterion after the Hume-Rothery rule $(r_i–r_{Fe})/r_{Fe}$, where the metallic radius of iron r_{Fe} = 126 pm; and activation energy of diffusion in в αFe.

Table 3. Properties of inhibitors.

Grain growth inhibitors	Solubility in αFe at 1000K, at% [14, 15]	Metallic radius r_i, pm [16]	Hume-Rothery rule $(r_i–r_{Fe})/r_{Fe}$, %	Activation energy for diffusion in αFe D, kJmol^{-1} [17]
Nb	0.12	146	15.9	299
W	1.0	139	10.3	–
Mo	2.5	139	10.3	282
V	unlimited	134	6.3	274
Cr	unlimited	128	1.6	267

All inhibitory elements in Table 3 are arranged in the order of increasing solubility in αFe. According to the Hume-Rothery rule [18], the solubility limit is reached when the atomic radius of the solute is by 15 % higher than that of the solvent. In accordance with this criterion, Nb atoms are virtually soluble in αFe, whereas the solubility of other elements rises in array Nb, W, Mo, V, Cr. With an increase in solubility of elements, the activation energy of diffusion consistently lowers.
The process of crystallization is accompanied by an increase in the volume of crystalline phase at the expense of migration of grain boundaries. Impurity (inhibitory) atoms interact with a moving boundary so that the boundary either carries these atoms with the velocity of its motion not exceeding that of atomic diffusion or drives off from them [19]. At a small velocity of the boundary migration, atoms with lower diffusion mobility, concentrating near the front of the moving boundary and remaining undissolved in the crystalline phase, provides a stronger drag.
Thus, the drag effect of the grain boundaries in the process of crystallization sequentially weakens in the array Nb, W, Mo, V, Cr, which manifests itself in the lowering of the crystallization-onset temperature T_x, as well as in the increase in the rate of elevating temperature V_{max} and in the peak temperature of the core heating T_{max}. The change in the velocity of migration of grain boundaries affects the extent of the crystallization process. Thus, the alloy with Nb has a structure characteristic of the initial stage of crystallization, namely, fine grains with the presence of a significant amount of grains 2 nm in size, and a predominant volume fraction of the αFe-Si solid solution. At the subsequent stages of crystallization, the grain size grows, the fraction of the stable Fe_3Si phase increases, and then, iron borides precipitate.

4. Summary

In the paper, the crystallization process of nanocrystaline alloys $Fe_{73.5}Cu_1M_3Si_{13.5}B_9$ where M = Nb, W, Mo, V, Cr and the physical origin of different efficiency of impact that inhibitors exert on the structure and magnetic properties has been investigated. It is shown that the structure formed at the peak temperature of the crystallization process is close to the optimal one and ensures high values of magnetic permeability. The efficiency of inhibitors is directly related to the solubility of inhibitory elements in αFe, which in turn affects the diffusivity of atoms. Upon slow migration of grain boundaries, the inhibitory atoms with a lower diffusivity, being concentrated near the front of moving boundary, provide a stronger drag. The weakening of the effect manifests itself in the lowering of the onset temperature of crystallization, as well as in the increase in the rate of elevating temperature and in the peak temperature of the core heating.

5. Acknowledgments

This work was supported by the Scientific researches of higher education institutions within the State task of the Russian Federation No. 4.9541.2017/8.9.

6. References

[1] Y Yoshizawa, *Handbook of Advanced Magnetic Materials*, Properties and Applications Vol. 4, ed D J Sellmyer and D Shindo, Springer (2006), pp. 124 – 158.
[2] Y Yoshizawa, S Oguma and K. Yamauchi J, *Appl. Phys.* 64 (1988), pp. 6044 – 6046.
[3] G Herzer, *IEEE Trans. Magn.* 25 (1989), pp. 3327 – 29.
[4] G Herzer, *Handbook of Magnetic Materials*, ed. K H J Buschow, 10 (1997), pp. 415 – 462.
[5] K Hono, D H Ping, M Onhuma and H Ondera, *Acta Mater.* 47 (1999), pp. 997 – 1006.
[6] W Lefebvre, S Morin-Grognet and F Danoix, *J. Magn. Magn. Mater.* 301 (2006) pp. 343 – 351.
[7] Y Yoshizawa and K Yamauchi, *Mater. Sci. Eng.* A133 (1991), pp. 176–179.
[8] M Müller and N Mattern, *J. Magn. Magn. Mater.* 136 (1994) 79–87.
[9] V A Peña Rodríguez, J F Regalado, E Baggio-Saitovitch and E C Passamani, *J. Alloys Comp.* 379 (2004), pp. 23–27.
[10] Wei Lu, Junwei Fan, Yuxin Wang and Biao Yan, *J. Magn. Magn. Mater.* 322 (2010), pp. 2935–37.

[11] J M. Borrego, C F Conde, A Conde and Phil. *Mag. Lett.* 80 (2000), pp. 359 – 365.
[12] J M Silveyra, E Illeková, P Švec, D Janičkovič, A Rosales-Rivera and V J Cremaschi, *Phys.* B 405 (2010), pp. 2720 – 25.
[13] S Tumanski, *Handbook of magnetic measurements*, CRC Press, Taylor&Francis Group (2011).
[14] O Kubaschewski, Iron–binary phase diagrams, Springer, 1982.
[15] Diagrams of binary metallic systems, 2, ed. N P Lyakishev, Moscow, 1997.
[16] N N Greenwood and A. Earnshaw, *Chemistry of the elements*, Butterworth-Heinemann, 1998.
[17] N Oono, H Nitta and Y Iijima, *Mater. Trans.* 44 (2003), pp. 2078–83.
[18] W Hume-Rothery and G V Raynor, *The structure of metals and alloys*, The institute of metals, 1962.
[19] G Gottstein and L S Shvindlerman, *Grain boundary migration in metals: thermodynamics, kinetics, applications*, CRC Press, 2010.

IOP Conf. Series: Materials Science and Engineering **380** (2018) 012012 doi:10.1088/1757-899X/380/1/012012

Synthesis and characterization of Co-Mo alloy coating

N R N Masdek[1], Z Salleh[1], M H Koay[2] and M A Ismail[1]

[1]Faculty of Mechanical Engineering, Universiti Teknologi MARA, Selangor, Malaysia
[2]Faculty of Mechanical Engineering, Universiti Teknologi MARA, Pulau Pinang, Malaysia

Abstract. Surface coating is very important in some applications. The preparation of Co-Mo alloy coating was prepared using electrodeposition process. The process of electrodeposition method was performed using 4 different concentration of molybdenum. The purpose of using different concentration of Mo in the solution bath is to find the optimum concentration of Mo that resulting to the best mechanical properties. The corrosion behavior, wear properties, surface morphologies and compositions of Co-Mo alloy was studied. The smooth surface was obtained when there was no Mo element in the solution bath. The Co-Mo coating that has the highest concentration of Mo, having the highest hardness value which is 286.6 HV. It was observed that the addition of Mo element in the deposit improved corrosion behavior of the deposits, hardness properties and resistance toward slurry erosion.

1. Introduction

Iron group commonly used as a source of magnetic materials [1]. The introduction of Mo elements in iron group material have the significant added value towards the magnetic properties [1]. Co-Mo alloy have the higher hardness and good corrosion resistance properties [1]. Co-Mo alloy coating have good strength, good corrosion resistance, good wear properties that can be applied to micro sensor, high force-force actuator and micro-electromechanical[2]. Surface coating is a technology that protects the surface of coated substrate from damage due to their application. Electrodeposition method is one of the finest techniques to produce Co-Mo alloy coating due to the low cost and simpler method to produce Co-Mo alloy coating [3]. Co-Mo alloy coating expected to be a good protective layer to improve wear, corrosion and hardness performance either in the mechanical, chemical and physical aspects. In this project, the wear properties of Co-Mo alloy coating were studied through the Slurry Erosion machine. The wear testing variable is the different concentration of Mo (33wt. %, 28wt. % and 23wt. %) at 500 rpm speed for 8 hours. The Co-Mo deposits were characterized through SEM and XRD. The corrosion behavior of the deposits was studied through Potentiodynamic polarization (PDP). The surface morphology of the Co-Mo deposits was analyzed using Surface measuring instrument (SURFTEST SV-600) machine was used to get the surface roughness. Finally, Vickers Microhardness tester was used to gain the hardness data of the Co-Mo alloy deposits.

2. Methodology

2.1. Sample preparation

Basically, there are two types of samples type 1 and type 2 as shown in Figure 1 which are 42.88cm2 and 59.2cm2 exposed areas respectively. Type 1 sample was made for the slurry erosion test that required maximum dimension of 2.54 cm x7.62 cm x 0.635 cm. Type 2 specimens was made for SEM, XRD, surftest and Vickers Microhardness testing. After the specimens cut into desired dimension, then the sample grinded with two different grit of sandpaper. Firstly, the sample was grinded with 180 grit sandpaper. After that, the samples grinded with 400 grit sandpaper. The purpose of grinding is to eliminate the galvanize layer at the surface of the samples.

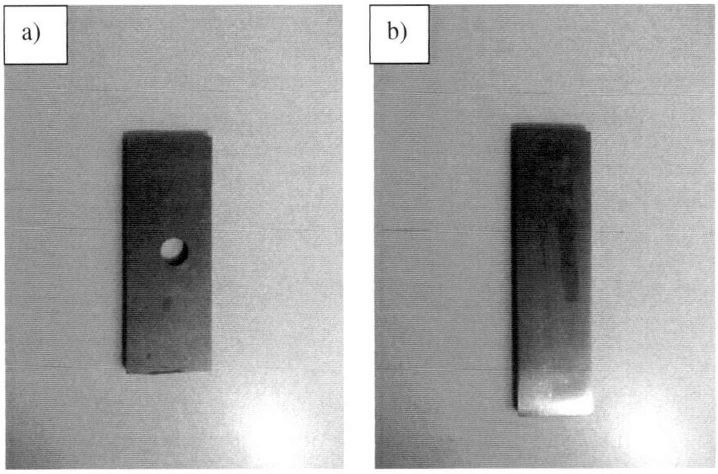

Figure 1. (a) Type 1 sample, and (b) Type 2 sample

2.2. Electrodeposition process

There are four type of solution with different concentration of Mo (0wt. %, 33wt. %, 28wt. % and 23wt. %) but with same concentration of Co for every solution that have been used while conducting the experiment. The preparation of the electrolytes bath has the same procedure. The magnetic stirrer will continuously stir the electrolyte bath until the end of the electrodeposition process. The temperature of the solutions is ±50°c. The ph. Value of the solution is 2.9. The boric acid was use as ph. buffer. The electrolyte is prepared in 500 ml distilled water. The current is around 0.9-1.0A and carried out for 30 minutes deposition time. Table 1 shows the chemical composition of solution used in electrodeposition process.

Table 1. The chemical composition of solution used in electrodeposition process

	Mole	Weight(g)
Cobalt Sulphate, $CoSO_4{\bullet}7H_2O$	0.02	7.02
Molybdenum, Mo	0.03	3.50
	0.02	2.80
	0.02	2.10
Boric acid, H_3BO_3	0.13	8.24
Ascorbic acid, $C_6H_8O_6$	0.03	5.86
Saccharine	0.01	1.50

2.3. Characterization techniques

The Co-Mo alloy coating was characterized using theX-ray diffraction (XRD) analysis. The scan region (2Θ) was range from 30° to 100° at rate scan of 4° min^{-1}. The morphology of the Co-Mo alloy deposits was studied using Scanning Electron Microscope (SEM). The SEM micrographs were taken with electron beam energy of 15keV.

2.4. Potentiodynamic polarization
The purpose of doing this testing is to obtain the corrosion behavior of the Co-Mo alloy coating. The specimens were attached to the electrical conductor wire using the wire glue. The purpose of this attachment is to enable the specimens fully immersed in the Phosphate Buffered Saline.

2.5. Vickers microhardness
Basically, Vickers Microhardness test is a testing to obtain the hardness values on the surface of the samples. The minimum and maximum load that available in the Vickers Microhardness machine is 10 g and 1 kg respectively. The load used for every sample is 1 kg. The reading was taken three times of the different spot of Co-Mo alloy coating sample to obtain average value.

2.6. Surface roughness
Surface roughness testing used to gain knowledge about the surface roughness of Co-Mo alloy coating on the samples. Surface roughness machine that available in the lab is Mitutoyo Surface measuring instrument (SURFTEST SV-600). The accurate procedure of this surface roughness testing is to take 3 different readings from 3 different spots on the sample. The purpose of taking the 3 different reading is to get the average value of the surface roughness value.

2.7. Slurry erosion
The specimen dimension for this slurry erosion wear test is 76 mm x 25 mm x 6 mm. Rotation speed of 500 rpm was for 8 hour with 2 hours' time interval for data collection. Mining sands with particles size of 710 μm were used in slurry erosion test.

3. Results and discussion

3.1. XRD analysis

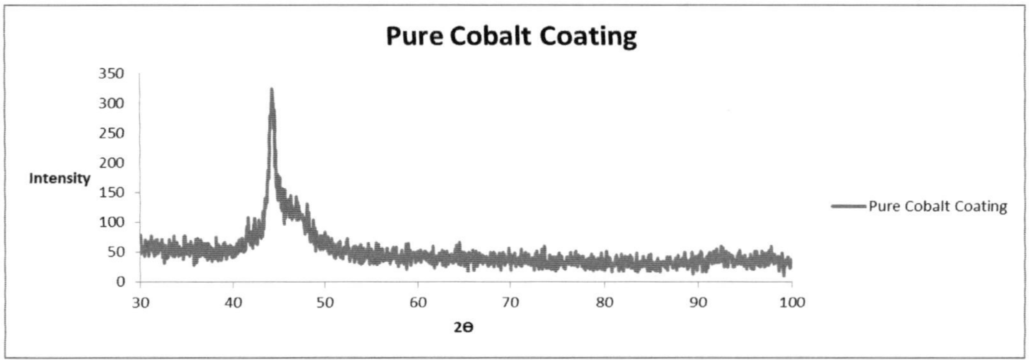

Figure 2. XRD pattern of pure cobalt coating.

The structural analysis of Co-Mo alloy deposits was performed using XRD. Based on the result obtained, the pure cobalt deposits and Co-Mo deposits show similar pattern of peak. Figure 3.1 –

IOP Conf. Series: Materials Science and Engineering **380** (2018) 012012 doi:10.1088/1757-899X/380/1/012012

Figure 3.4 shows the result of the XRD obtained from the Co-Mo alloy deposits. Pure Co and Co-Mo coating samples showed the peaks of Co and Co-Mo after referring to the theta values with the XRD library data for Co-Mo. The pure Co deposits rich with close-packed hexagonal structure [4]. The only phase detected in pure Co coating is close-packed hexagonal structures (hcp) which are hcp (100) at $2\theta = 41.58^o$, hcp (002) at $2\theta = 44.50^o$, hcp (101) at $2\theta = 47.44^o$ and hcp (112) at $2\theta = 92.44^o$. For the Co-Mo deposits (23-33 wt. %), the diffraction peaks of Mo at $2\theta = 40.51\ ^o$, $58.60\ ^o$ and $73.66\ ^o$ correspond to (110), (200) and (211) plane structure. However, the diffraction of Mo peaks very hard to obtain using electrodeposition method. The more content of Mo in the Co-Mo deposits shift the peaks of Mo towards the lower diffraction [5].

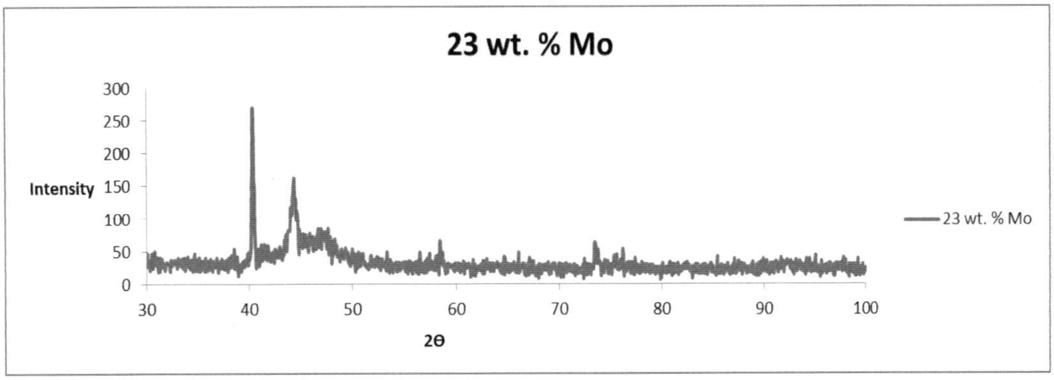

Figure 3. XRD pattern of Co-Mo alloy coating with 23wt. % Mo.

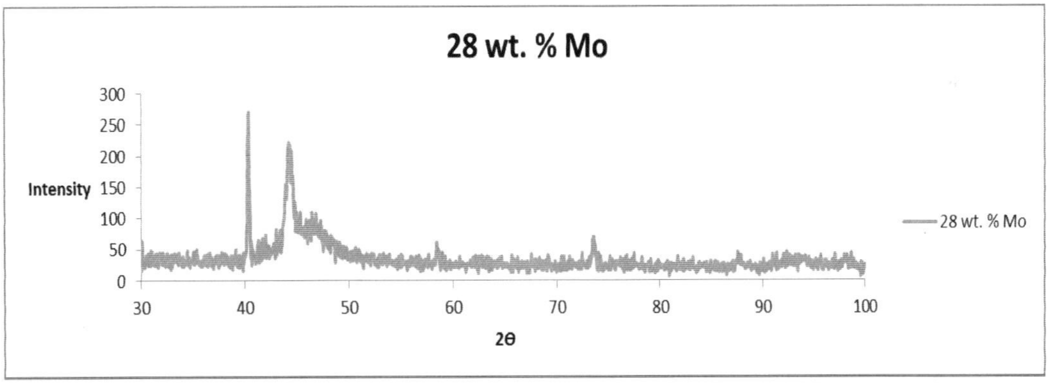

Figure 4. XRD pattern of Co-Mo alloy coating with 28wt. % Mo.

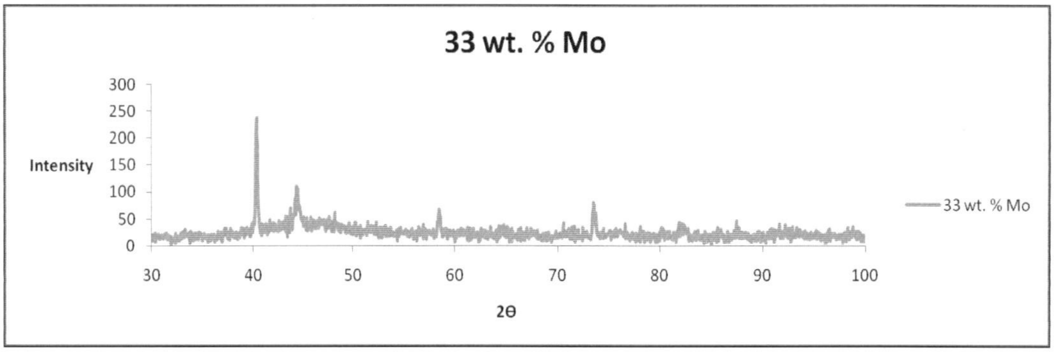

Figure 5. XRD pattern of Co-Mo alloy coating with 33wt. % Mo.

3.2. Scanning electron microscopy (SEM)

Figure 3.5 shows the micrograph of pure Co and Co-Mo deposits obtained through SEM. Based on the SEM results, the Co-Mo deposits that content the highest concentration of Mo (33 wt. % Mo) show finest grain and crack free surface morphology compared with the other Co-Mo deposits. Besides, the particles distribution of Co-Mo deposits in the highest Mo content seemed to be uniform distribution compared to the other Co-Mo deposits. The Co-Mo deposits with the lowest Mo content show poor distribution of particles. The Co-Mo deposits with the moderate concentration of Mo (28 wt. %) show very coarse surface morphology. The pure Co deposits show cracks on the surface morphology while the surface morphologies after adding Mo are crack-free. These results are consistent with previous research [6].

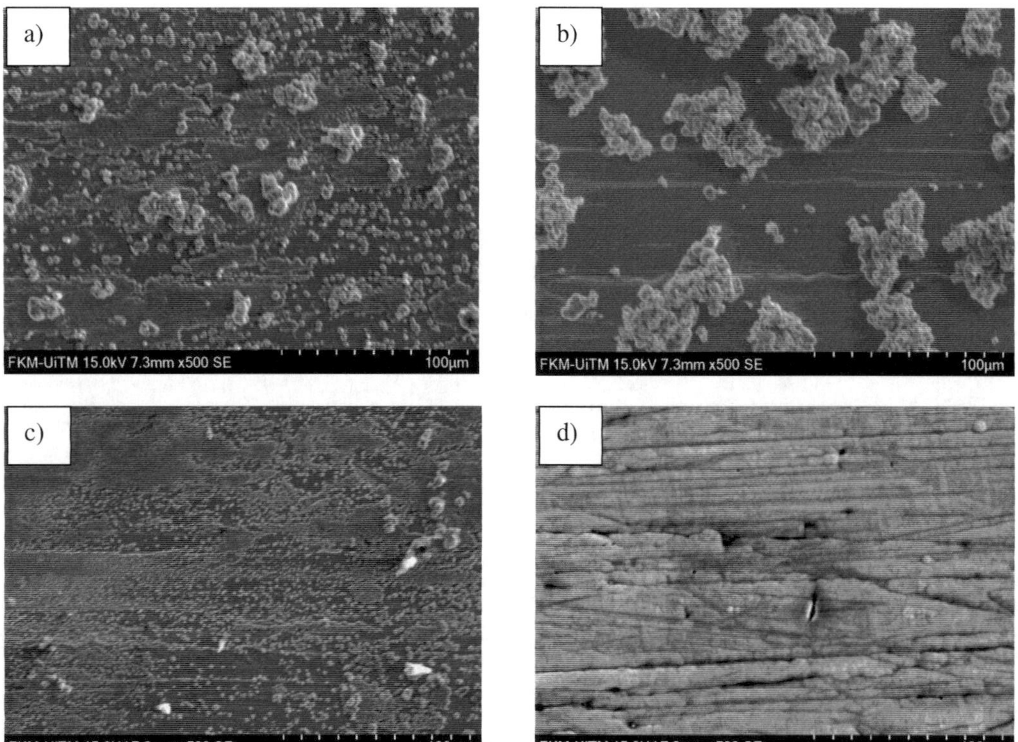

Figure 6. (a) 23 wt. % Mo, (b) 28 wt. % Mo, (c) 33 wt. % Mo, (d) Pure Co

3.3. Vickers microhardness

The Microhardness of the samples was taken for 3 times at different spots of the deposits to ensure the average reading were taken. Figure 3.6 show the Vickers Microhardness value of pure Co deposits and Co-Mo deposits with different concentrations of Mo. Based on the Vickers Microhardness results, the highest Microhardness value was obtained at 33 wt. % Mo while the lowest Microhardness value was obtain at pure Co deposit. The Co-Mo deposits with the lowest Mo content show the second highest of Microhardness value. The highest value of Microhardness obtained in 33 wt. % Mo was due to the finer surface morphology as shown in previous SEM sub-topic. The fine surface morphology improved the value of Microhardness [4]. The Co-Mo deposits have good hardness properties depending on the surface morphology and the concentration of Mo content in the deposits [7] [8].

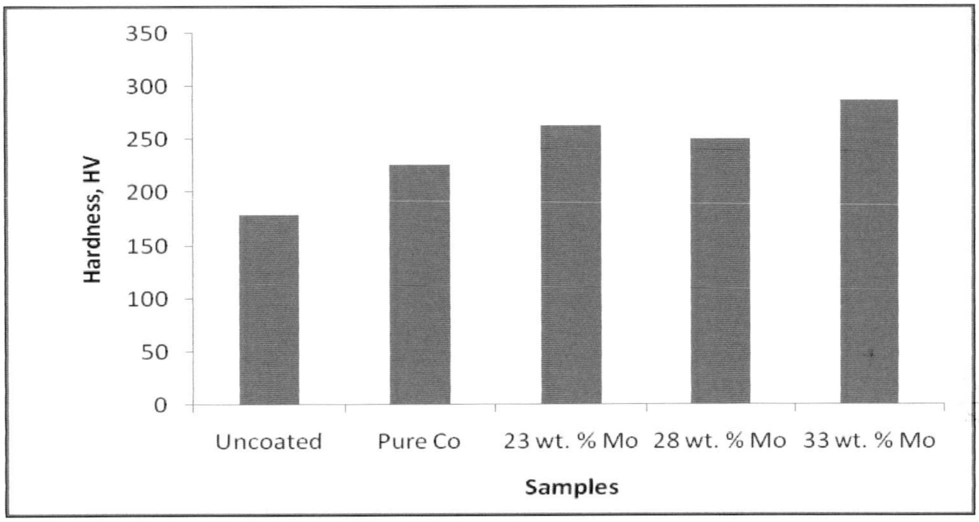

Figure 7. Vickers Microhardness value for different concentrations of Mo.

3.4. Potentiodynamic polarization

Figure 3.7 show the Potentiodynamic polarization (PDP) curves of different concentrations of Mo. The Co-Mo deposits that contain the highest concentration of Mo show the best corrosion resistance. This is shown as the corrosion potential of 33 wt. % Mo is the highest and the corrosion current is the lowest as shown in Table 3.1. This is because of the formation of thin passive film which increases the corrosion resistance [9]. The pure Co deposits show the poorest corrosion resistance as corrosion potential is the lowest. The corrosion rate of the Co-Mo deposits decrease as the molybdenum content increase [2]. However, the corrosion resistance of Co-Mo deposits in 23 wt. % Mo is better than in 28wt. %. This can be suggested that, the addition of Mo element in the deposits gave the significant added value towards corrosion resistance, but at the same time the surface morphology also play the main role in the corrosion properties [2] Co-Mo deposits that content 28 wt. % Mo show the worst corrosion resistance compared to deposits with 23 wt. % Mo resulting from the coarsest surface

IOP Conf. Series: Materials Science and Engineering **380** (2018) 012012 doi:10.1088/1757-899X/380/1/012012

morphology as shown in SEM. However, Co-Mo alloy coating have the good corrosion resistance depending on the concentration of Mo content and surface morphology [10].

3.5. Surface roughness

Figure 3.8 show the average surface roughness of different concentrations of Mo. The reading of the surface roughness was taken for three times at different spots to get the average values of surface roughness of the deposits. It was observed that, the surface roughness was varied for different concentrations of Mo. The pure Co deposits show the lowest surface roughness (Ra = 0.306) due to its finest surface morphology and uniform particles distribution. Although the pure Co deposits show the good surface roughness, however the surface also shows the crack films. Meanwhile, out of all of the deposits with Mo, the Co-Mo deposits (33 wt. % Mo) show the lowest surface roughness. This is because the surface morphology for this particular deposit is the finest as shown in SEM. Co-Mo deposits with the moderate content of Mo (28 wt. %) show the poorest surface roughness result (Ra = 6.47). This is due to the ununiformed distribution of particles and very coarse grains as shown in SEM. The surfaces structure of the deposits affected the result of the surface roughness of the deposits. The finest surface morphology improved the surface roughness value.

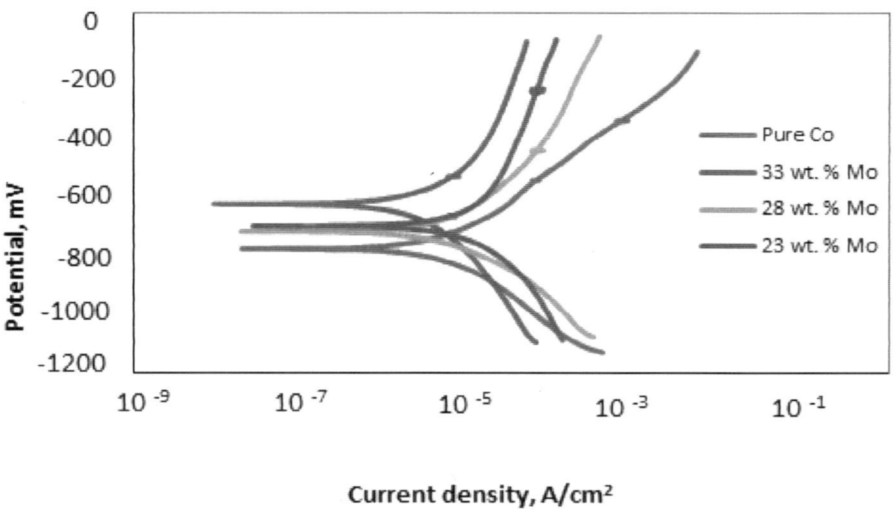

Figure 8. Potentiodynamic polarization curves of different concentrations of Mo.

3.6. Surface roughness

Figure 3.8 show the average surface roughness of different concentrations of Mo. The reading of the surface roughness was taken for three times at different spots to get the average values of surface roughness of the deposits. It was observed that, the surface roughness was varied for different concentrations of Mo. The pure Co deposits show the lowest surface roughness (Ra = 0.306) due to its finest surface morphology and uniform particles distribution. Although the pure Co deposits show the good surface roughness, however the surface also shows the crack films. Meanwhile, out of all of the deposits with Mo, the Co-Mo deposits (33 wt. % Mo) show the lowest surface roughness. This is because the surface morphology for this particular deposit is the finest as shown in SEM. Co-Mo deposits with the moderate content of Mo (28 wt. %) show the poorest surface roughness result (Ra = 6.47). This is due to the ununiformed distribution of particles and very coarse grains as shown in SEM. The surfaces structure of the deposits affected the result of the surface roughness of the deposits. The finest surface morphology improved the surface roughness value.

3.7. Slurry erosion

Figure 3.9 show the result the slurry wear test for different concentrations of Mo. The moderate rotation speed used for the slurry erosion test was 500 rpm. The moderate rotation speed was used in order to get the most optimum concentration of Mo with the best wear properties. The erosion wear rate (mass loss of deposits) increases with time [11]. The result of slurry erosion test shows the similar trend of curve as previous research [12]. The result from the slurry wear test shows that, the deposits with the highest amount of Mo content (33 wt. % Mo) show the least mass loss which is 0.0833g. The pure cobalt also shows great wear properties as the mass loss of deposits is 0.0861g. The mass loss of deposits through the slurry erosion test is related to the surface roughness of the deposits. The poorest surface morphology (28 wt. % Mo) show the most mass loss of deposits through slurry erosion test. The finest surface roughness tends to good in terms of wear properties. However, the addition of Mo into the Co based deposits gave the significant added value toward the wear properties [13].

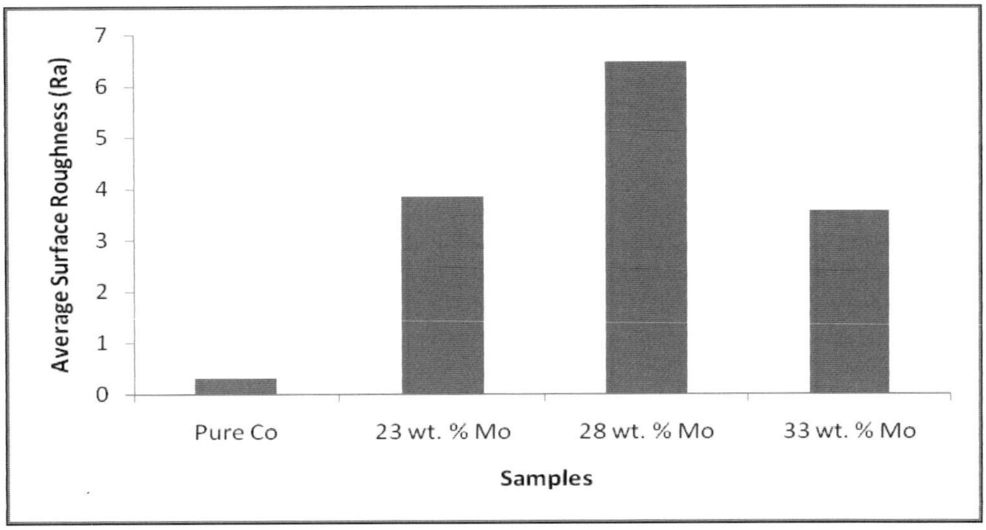

Figure 9. Average surface roughness for different concentrations of Mo.

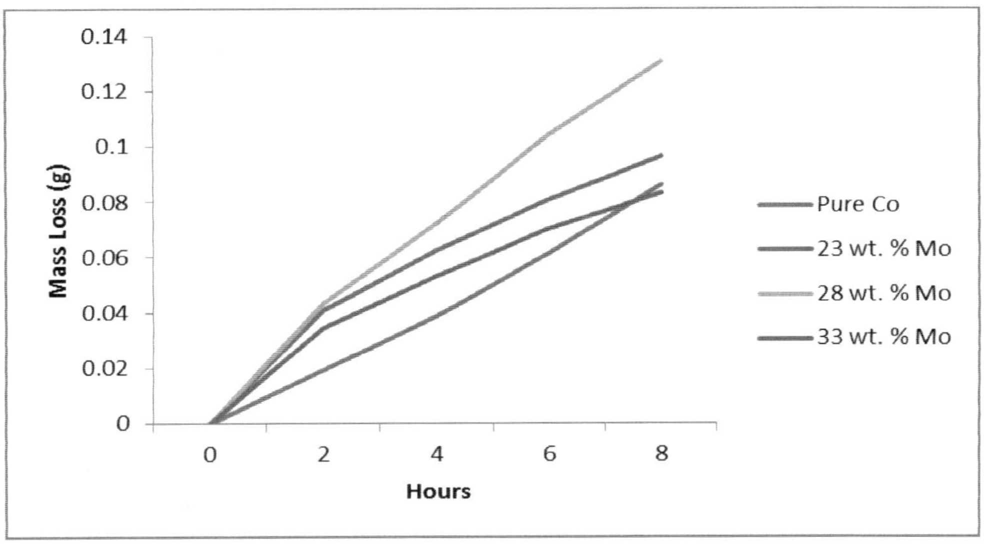

IOP Conf. Series: Materials Science and Engineering **380** (2018) 012012 doi:10.1088/1757-899X/380/1/012012

Figure 10. Slurry wear test comparison for different concentrations of Mo.

4. Conclusions

Based on the result obtained from the study, the addition of Mo gave the significant added value toward the surface morphology, corrosion resistance, hardness and wear properties of Co-Mo alloy coatings. The Co-Mo alloy deposits with the highest Mo content (33 wt. % Mo) show fine surface morphology obtained from SEM resulting good corrosion resistance, good hardness properties and good wear resistance.

The surface coating of Co-Mo alloy improved the actual hardness of the uncoated mild steel. The actual Vickers Microhardness of the uncoated mild steel is 179.67 HV. As the Co-Mo alloy coating was coated the mild steel, the Vickers Microhardness values was increased as the increased of Mo content in the Co-Mo coating. The highest Vickers Microhardness was obtained in 33 wt. % Mo which is 286.6 HV. Surface roughness of every Co and Co-Mo deposits was different according to the Mo content in the deposits. The pure Co deposit was the lowest surface roughness of 0.306 Ra while the 28 wt. % Mo deposits was the highest surface roughness of 6.47 Ra.

The Co-Mo deposits that contain the highest concentration of Mo show the best corrosion resistance. This is shown as the corrosion potential of 33 wt. % Mo is the highest and the corrosion current is the lowest. The corrosion rate of the Co-Mo deposits decrease as the molybdenum content increase. However, the surface morphology also affects the corrosion behavior of Co-Mo deposits. Least coating mass loss was obtained in the 33 wt. % Mo deposits while the highest coating mass loss was obtained at in the 28 wt. % Mo deposits. This is due to the finest surface morphology obtained in 33 wt. % Mo that improved the wear resistance of the deposits.

Acknowledgments

The authors gratefully acknowledge the help of the Research Management Centre (RMC) and Ministry of Higher Education (MOHE) Malaysia in providing the Fundamental Research Grant Scheme (Project Number: 600-RMI/FRGS 5/3 (80/2015) grants and Universiti Teknologi MARA for their support.

References

[1] Zhou Lu L, Yu L, Xu X and Y. Jiang *Electrochimica Acta* 2013 **106** pp 258–263
[2] Kublanovskii V, Yapontseva Y, Troshchenkov Y and Gromova V *Russian Journal of Applied Chemistry* 2010 **83** pp 440–444
[3] Marita Y and Yaacob I *Advanced Material Research* 2010 **97** pp 1360-63
[4] Srivastava M, Yoganandan G and Grips V *Surface Engineering* 2012 **28** pp 424–429
[5] Gómez E, Pellicer E, Alcobé X and Vallés E *Journal of Solid State Electrochemistry* 2004 **8** pp 497–504
[6] Gómez E, Pellicer E and Vallés E *Journal of Electroanalytical Chemistry* 2001 **517** pp 109–116
[7] Beltowska-Lehman E and Indyka P *Thin solid films* 2012 **520** pp 2046–51
[8] Han Q, Cui S, Pu N, Chen J, Liu K and Wei X *International Journal of Hydrogen Energy* 2010 **35** pp 5194–5201
[9] Anicai L, Costovici S, Cojocaru A, Manea A and Visan T *Transactions of the Institute of Metal Finishing* 2015 **93**
[10] Jakˇsi M *International Journal of Hydrogen Energy* 1987 **12** pp 727–752
[11] Dhindsa, Dalbir S and Mohapatra S 2011 *Parametric study of erosion wear of solid-liquid mixture* ***Thesis*** Master of Engineering, Thapar University India
[12] Patel S and Sejkar S *International Journal of Eengineering & Research* 2014 **3** pp 2277-5668
[13] Malayoglu U and Neville A *Wear* 2005 **259** pp 219–229

IOP Conf. Series: Materials Science and Engineering **380** (2018) 012012 doi:10.1088/1757-899X/380/1/012012

First name	Surname	Institution	Email address	Title (Prof/ Dr./student)
Nik Rozlin	Nik Masdek	Universiti Teknologi MARA	nikrozlin@yahoo.com	Dr.
Zuraidah	Salleh	Universiti Teknologi MARA	szuraidah@salan.uitm.edu.my	Dr.
Mei Hyie	Koay	Universiti Teknologi MARA	koay@ppinang.uitm.edu.my	Dr.
Mohammad Adam	Ismail	Universiti Teknologi MARA	adamismail_93@yahoo.com	Mr.

IOP Conf. Series: Materials Science and Engineering **380** (2018) 012013 doi:10.1088/1757-899X/380/1/012013

The effect of structural changes on the mechanical properties of the Mg-1%Zn-0.2%Ca alloy processed by ECAP

O B Kulyasova[1,2], R K Islamgaliev[1]

[1] Institute of Physics of Advanced Materials, Ufa State Aviation Technical University, 12 Karl Marx Street, Ufa, 450008, Russia

[2] Saint Petersburg State University, 7/9 Universitetskaya nab., St. Petersburg, 199034, Russia

E-mail: elokbox@mail.ru

Abstract. The structure of the magnesium alloy Mg-1%Zn-0.2%Ca after equal-channel angular pressing (ECAP) has been studied by optical and scanning electron microscopy. It was established that increasing the equivalent strain by increase a number of passes through a ECAP die leads to more pronounced grain refinement in the alloy.When equivalent strain achieved 7.2, an average grain size of 2 μm was observed. It was shown that the ECAP samples enable to demonstrate the high microhardness of 69.5 HV and the enhanced ultimate tensile strength of 282 MPa, which is more than 2 times higher in comparison with the homogenized state, while maintaining ductility.

1. Introduction

The creation of new biodegradable materials is one of the priority directions in the development of materials science for the needs of medicine. As a result a durable implant degrading with a controlled speed, which does not have a harmful effect on the human body [1-3] and performs its functions during the time needed to repair the damaged bone (12-14 weeks) has been developed. Magnesium, which can be used in biodegradable implants, is of special attention by specialists, because it is completely biocompatible[1].

The difficulty is that pure magnesium is unstable and dissolves very quickly in the high chloride environment of the physiological system, thereby losing its functions before the bone is fully restored. Increasing the strength characteristics of magnesium can be achieved due to of a large number of solutes. However, because of the solubility of the alloy in the human body, the choice of alloying elements is significantly narrowed, they must be biologically inert[4,5].

Moreover, it was found that as a result of a high content of alloying elements, secondary phases are formed in the alloys, which are barriers to the movement of dislocations during plastic deformation and, accordingly, the ductility sharply decreases in the alloy.Thus, a weakly alloyed Mg-1%Zn-0.2%Ca alloy was selected in which the volume fraction of the secondary phases isminimized.

The selected alloying chemical elements are not toxic to the body and do not cause negative consequences [6-9]. In addition it is known that grain refinement can to be used to increase the strength of various metals and alloys, including Mg alloys [10-13]. At the same time for application in medicine it is important to produce bulk ultrafine-grained materials, for example, by equal channel angular pressing (ECAP). However there are only limited publications about microstructure and properties of the Mg-Zn-Ca alloys processed by ECAP [14]. Therefore, the purpose of this work was to identify the evolution of the structure in the process of ECAP and its impact on mechanical properties of the Mg- 1%Zn-0.2%Ca.

2. Experimental

As the initial material for the studies, the cast Mg-1%Zn-0.2%Ca alloy was chosen. The initial cast samples were subjected to homogenization annealing at a temperature of 450° C for 24 hours with cooling into water. Heat treatment of the samples was carried out in a muffle furnace Nabertherm. The refinement of the structure was carried out by equal-channel angular pressing (ECAP). ECAP was carried out in a die-set with a channels intersection angle of 120° at a speed of 6 mm/min on samples with a diameter of 20 mm after preheating the billet at a pressing temperature of 20 minutes.

To reveal the evolution of the structure, several ECAP regimes were tried. In case of regime#1, two pressing passes were carried out at a temperature of 400°C. Incase of regime#2, two passes were added at a temperature of 350°C to the existing two passes at a temperature of 400°C. Under regime#3 to the available two passes at a temperature of 400°C, two passes at a temperature of 350°C were added two more passes at a temperature of 300° C. In case of regime #4, two passes were added at a temperature of 250°C to two passes at a temperature of 400° C, two passes at a temperature of 350°C, and two passes at a temperature of 300° C. The equivalent strain in case of regimes # 1, 2, 3 and 4 was 1.8,3.6,
5.4 and 7.2, respectively.

The macrostructure was examined with an Olympus GX51 optical microscope. The microstructure analysis was performed with a JEM-6390 scanning electron microscope (SEM) with accelerating voltages of 10 kV. The microhardness (HV) was measured by the Vickers method on a Micromet 5101 measurement tool with a load of 50 g and a dwell time of 10 s. Tensile tests were performed on an Instron 5982 testing machine at room temperature and initial strain rate of $10^{-3}c^{-1}$ using specimens with a gage dimension of 0.6×1×4.5 mm^3. For each condition, a minimum of 5 samples was tested.

3. Results anddiscussion

The structure of the Mg-1%Zn-0.2%Ca alloy after homogenization annealing consisted of equiaxed grains of α-Mg with an average size of 270 μm (Fig.1a). In the body of the grains there were also Mg_2Ca particles up to 1 μm in size, and at the boundary the size of these particles was 4 μm, the total volume fraction of Mg_2Ca particles in the Mg-1%Zn-0.2%Ca alloy samples was 4%. The annealing twins of various sizes were clearly seen in the structure of all the investigated samples (Fig.1b).

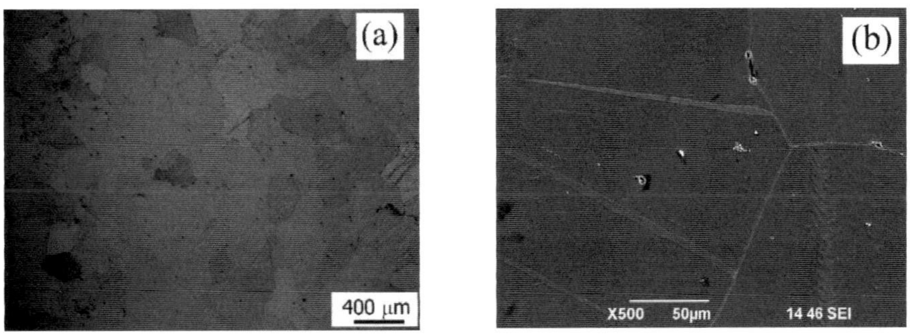

Figure 1. Structure of magnesium alloy Mg-1%Zn-0.2% Ca in the initial state:
a - optical microscopy; b - scanning electron microscopy.

After two ECAP passes one observes the grain refinement (Fig. 2a). However, the structure is highly heterogeneous in volume. Approximately 40% of the structure is occupied by regions with elongated grains. In there maining volume, equiaxed grains with an average size of 5.1μm were formed. Particles in the structure were not detected. A possible and most probable reason may be their refinement and transformation into a finely dispersed form. With an increase in the number of ECAP passes to four,an even finer structure with an average grain size of approximately 4.0 μm is formed (Fig. 2b). However, the heterogeneity is still observed, because separate areas with elongated grains are clearly seen. More homogeneous structure was formed with the number of ECAP passes equal tov6.The average grainbsize was 3.7 μm (Fig. 2c). The smallest average grain size of 2 μm and a homogeneous structure (Fig. 2d) was produced after a total number of passes equal to 8.

The microhardness of the alloy in the homogenized coarse-grained samples was 42.3 ± 4.8 HV. Structural changes in the ECAP process led to an increase in the microhardness. Thus, after two ECAP passes, the microhardness is increased to 49.6 ± 5.6 HV, and after 4, 6 and 8 passes to 59.5 ± 5.8, 63.1±6.3 and 69.5±6,7HV, respectively (Fig.3). Obviously, the greater is the degree of strain, the

IOP Conf. Series: Materials Science and Engineering **380** (2018) 012013 doi:10.1088/1757-899X/380/1/012013

smaller is the average grain size and the higher is the microhardness. Accordingly, the highest values of microhardness (69.5 ± 6.7 HV) were achieved at the highest equivalent strain (7.2) and the smallest average grain size (2μm).

Figure 4 shows the tensile curves of the alloy under study for different ECAP samples. The Mg-1%Zn-0.2%Ca alloy, in the homogenized state has a higher strength (138 MPa) than pure magnesium (35 MPa [14]). In the deformed state, its properties improved gradually, as expected, with an increase in the equivalent strain. So, after two passes the ultimate tensile strength was only 150 MPa, after four and six passes it reached the values of 205 and 240MPa, and after eight ECAP passes increased to 282 MPa, which is two times higher than in the initial state.

Also, the yield stress was significantly increased from 34 MPa in the initial state to 205 MPa after deformation by ECAP with a total number of passes equal to 8. Based on the results obtained, it can be concluded that the average grain size significantly influences the strength of the alloy - the smaller is the average grain size, the higher is the strength.

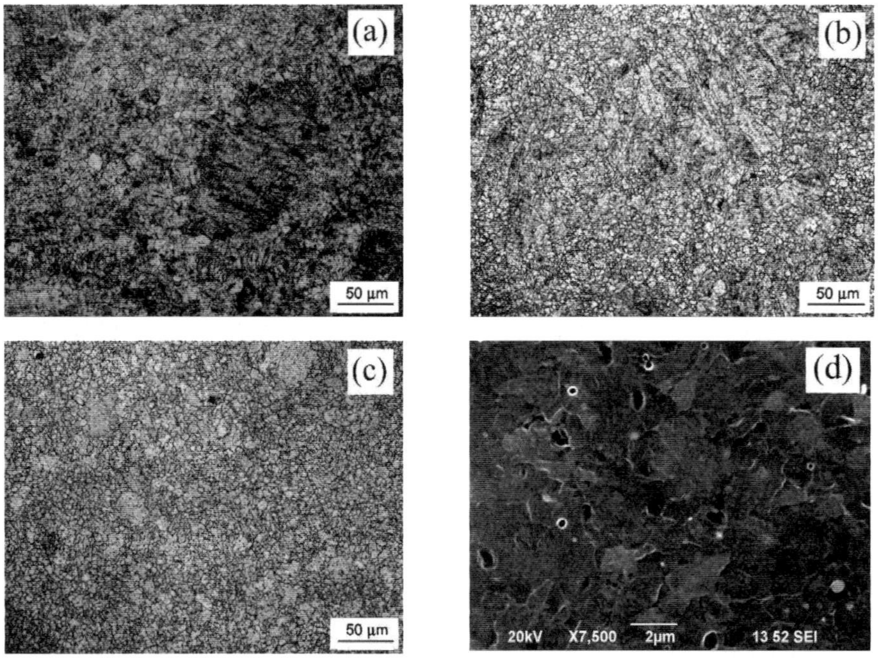

Figure 2. Structure of the samples of Mg-1%Zn-0.2% Ca alloys after ECAP:
a - 2 passes, b - 4 passes; c -6 passes; d - 8 passes.

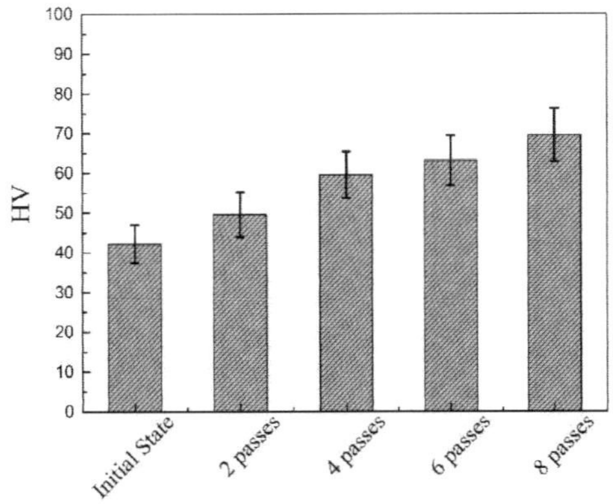

Figure 3. Microhardness of the Mg-1%Zn-0.2% Ca alloy in the initial state and subjected toECAP with a different number ofpasses.

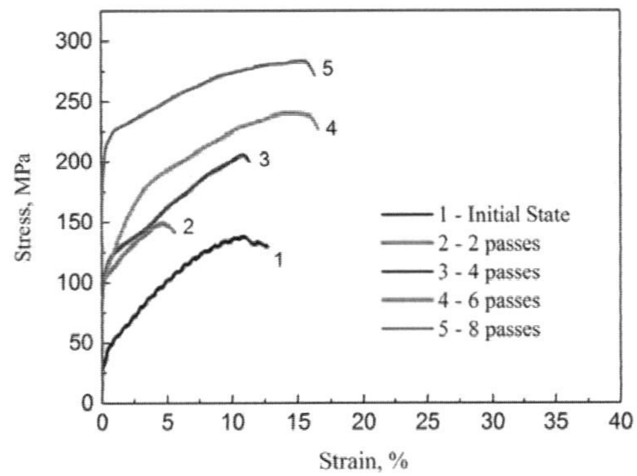

Figure 4. Tensile tests of the Mg-1%Zn-0.2%Ca alloy in the initial state and subjected to anECAP with a different number ofpasses.

According to the literature data the increase in the strength of magnesium alloys is implemented via refinement of the average grain size. However, in comparison with the known data [15], such high values of the ultimate tensile strength for this alloy were obtained for the first that is attractive for application in medicine to improve the design of implants. It can also be seen from the tensile curve that after two and four ECAP passes, a decrease in the elongation to failure is observed. This effect is probably associated with the formation in the structure of areas with elongated grains that embrittle the material. As the number of passes increases, the ductility increases slightly due to the formation of a homogeneous structure throughout the sample volume.

4. Conclusion

The microstructure and mechanical properties of the Mg-1%Zn-0.2%Ca alloy, promising for use in biodegradable implants after deformation by the equal-channel angular pressing, were investigated. It was shown that an increase in the equivalent strain led to the formation of a more homogeneous structure with a smaller average grain size.

The smallest average grain size (2 μm) was achieved at the highest equivalent strain (7.2). The structural parameters obtained during this deformation provided the greatest increase in the ultimate tensile strength up to 282 MPa, which is two times higher than the homogenized sample (138 MPa).At the same time a ductility did not deteriorate that is of great interest and makes this alloy promising for

IOP Conf. Series: Materials Science and Engineering **380** (2018) 012013 doi:10.1088/1757-899X/380/1/012013

its use as a material for manufacturing bioresorbable implants.

Acknowledgments
The work was supported by the Russian Foundation for Basic Research (No. 16.7268.2017), and St.Petersburg grant 6.37.204.2016. Research part of work has been executed with use of the equipment of the center for collective use "Nanotech", Ufa State Aviation Technical University.

References
[1] Peng T and Xuanyong L 2015 *Regenerative Biomaterials* **2** 135
[2] Staiger M P, Pietak A M, Huadmai J and Dias G 2006 *Biomaterials* **27** 1728
[3] Cha P R, Han H S, Yang G F, Kim Y C, Hong K H, Lee S C and Byun J Y 2013 *Scientific Reports* **3** 28 2367
[4] Bakhsheshi-Rad H, Reza H, Idris M H, Kadir M R, Farahany S, Fereidouni A and Yahya M 2012 *Applied Mechanics and Materials* **121** 568
[5] Nowosielski R 2013 *Journal of Achievements in Materials and Manufacturing Engineering* **61** (2) 367
[6] Zhang Y N, Liu X D, Altounian Z and Medraj M 2013 *Scripta Materialia* **68** (8) 647
[7] Guo Y B and Salahshoor M 2012 *Materials* **5** 135
[8] Bakhsheshi-Rad H R, Idris M H, Abdul-Kadir M R, Ourdjini A, Medraj M, Daroonparvar M and Hamzah E 2014 *Materials and Design* **53** 283
[9] Kubok K, Litynska-Dobrzynska L, Wojewoda-Budka J, Góral A and Debski A 2013 *Archives of Metallurgy and. Materials* **58** (2) 329
[10] Valiev R Z, Islamgaliev R K and Alexandrov I V 2000 *Progress Materials Science* **45** (2) 103
[11] Kulyasova O B, Islamgaliev R K, Zhao Y and Valiev R Z 2015 *Advanced Engineering Materials* **17** (12)
[12] Cizek J, Prochazka I, Smola B, Stulikova I, Očenášek V, Islamgaliev R K and Kulyasova O 2010 *Materials Science Forum* **633-634** 353
[13] Kulyasova O B, Islamgaliev R K, Kil'mametov A R and Valiev R Z 2006 *Physics of Metals and Metallography* **101** (6) 585
[14] Vinogradov A Yu, Vasil'ev E V, Linderov M L, Merson D L and Rjevskaja O E 2015 *Vector Science TSU* **34** (4) 18
[15] Hofstetter J, Becker M, Martinelli E, Weinberg A M, Mingler B, Kilian H, Pogatscher S, Uggowitzer P J, and Löffler J F 2014 *JOM* **66** (4) 566

IOP Conf. Series: Materials Science and Engineering **380** (2018) 012014 doi:10.1088/1757-899X/380/1/012014

Microstructure, mechanical and corrosion properties of ultrafine-grained Mg-2%Sr alloy

O.B. Kulyasova[1,2], R.K. Islamgaliev[1], E V Parfenov[3], Y.F. Zheng[4], R.Z. Valiev[1,2]

[1] Institute of Physics of Advanced Materials, Ufa State Aviation Technical University, 12 Karl Marx Street, Ufa, 450008, Russia
[2] Saint Petersburg State University, 7/9 Universitetskaya nab., St. Petersburg, 199034, Russia
[3] Department of Theoretical Basis of Electrical Engineering, Ufa State Aviation Technical University, 12 Karl Marx Street, Ufa, 450008, Russia
[4] College of Engineering, Peking University, 5 Yi-He-Yuan Road, Hai-Dian District, Beijing, China.

E-mail: elokbox@mail.ru

Abstract. Thermal stability of microstructure of the magnesium alloy Mg-2%Sr processed by high pressure torsion has been studied using scanning and transmission electron microscopy. It was shown that formation of bimodal structure after additional annealing at a temperature of 200°C leads to demonstration of enhanced ultimate tensile strength (245 MPa) with a ductility of 1.5%. In addition a uniform distribution of eutectic phase in bimodal structure led to a slowing down of the corrosion rate to 2% for 31 days (compared to the initial state of 78% for the same period).

1. Introduction

Many scientists are engaged in improving the biological properties of materials for the manufacture of medical implants [1-3]. Among the various materials, great attention is paid to metal implants. Magnesium and magnesium alloys have a great potential among other metals, because they are light and bioresorbable, and can be used as a material for the manufacture of screws, pins and plates should, if they have a high level of mechanical properties and a low rate of corrosion. Pure magnesium does not meet such requirements in the first place due to its low strength. Thus, for the effective use of magnesium as a biomaterial, it is necessary to increase its strength characteristics by alloying with other elements. However, the choice of alloying elements is not so wide, because these elements must also be biocompatible and not cause allergic reactions [4]. In the present work, Sr was chosen as the alloying element, since it is a natural element for the bone [5]. It was found out that the Sr content of 2wt% is optimal in the ratio of strength and ductility characteristics, but the cast Mg-2%Sr alloy still does not have sufficient mechanical properties to make medical products from it [5]. It is known that deformed magnesium alloys have a higher level of mechanical properties compared to cast alloys [7-9]. Recently, methods of controlling the structure have been actively developed by grain refinement and creating a uniform distribution of excess phases, based on the use of severe plastic deformation [10]. With the help of such methods, and in particular of high pressure torsion (HPT), deep structural changes occur, including significant grain refinement and partial dissolution of the excess phase components. In the present work we investigated the effect of high pressure torsion and additional thermal treatment on the structural-phase, mechanical and corrosion changes of magnesium alloy Mg-2wt%Sr.

2. Experimental

As the initial material, we used samples of an Mg-2wt%Sr alloy subjected to solid solution treatment at 430°C for 24 h and subsequent quenching.

To obtain a nanostructured state in a magnesium alloy Mg-2% Sr, disks with a diameter of 20 mm and a thickness of 1 mm were subjected to HPT at a pressure of 6 GPa at room temperature. For the development of the structure, 10 revolutions were performed with a strain rate of 1 rpm.

The macrostructure was studied with an Olympus GX51 optical microscope. The microstructure was examined with a JEM-6390 scanning electron microscope (SEM) and a transmission electron microscope (TEM) JEM-2100 with accelerating voltages of 10 kV and 200 kV, respectively. The discs with a diameter of 3 mm for TEM foils were cut out after grinding to 0.1 mm. Then they were subjected to twin-jet electropolishing on Tenupole-5. Electrolyte for electropolishing consisted of of HCl acid - 10% and methanol - 90%. The average grain size was determined by the intercept method and was calculated from more than 250 grains measurement.

The microhardness of samples HV was measured by the Vickers method using a Micromet-5101 device with a load of 1 N and a dwell time of 10 s. Each sample was measured along a diameter more than 20 times to provide reliable results.

Tensile tests were performed on an Instron 5982 testing machine at room temperature and initial strain rate of $10^{-3}c^{-1}$ using specimens with a gage dimension of $0.6\times1\times4.5$ mm^3. A set of 3 samples was tested for each condition.

The corrosion tests were carried out according to ASTM G1-03-E. This study presents the mass loss results of magnesium alloy in a 0,9% NaCl at temperature 37 °C.

3. Results and discussion

The structure of the Mg-2%Sr alloy after homogenization is a solid solution (α-Mg), with the eutectic (α-Mg + $Mg_{17}Sr_2$) located along the grain boundaries (Fig. 1). The average grain size was 252 ± 42 μm. The volume fraction of the eutectic was 17 ± 1%.

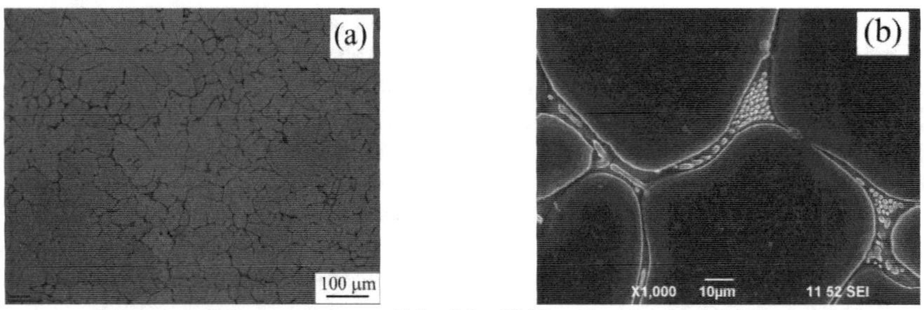

Figure 1. Microstructure of the Mg-2%Sr alloy in the initial samples

As a result of high pressure torsion, an ultrafine-grained structure with an average grain size of 432 ± 52 nm was formed in the alloy (Fig. 2). Investigations of microstructure of the HPT samples revealed a high density of dislocations in both the overall microstructure (Fig. 2a) and individual grains (Fig. 2b). However, small recrystallized grains with a size of about 200 ± 24 nm, free from dislocations (Fig. 2c) were also found out. A significant difference in the structure of the eutectic (Fig. 2c) was found in comparison with the initial state. SEM images showed the eutectic phase is noticeably partitioned, and its volume fraction decreased to 11 ± 1%, which indicates its partial dissolution at plastic deformation.

In the microstructure of the samples subjected to the HPT and additional annealing at a temperature of 150 °C, a small increase in the average grain size to 530 ± 37 nm was observed due to migration of grain boundaries (Fig. 3a). Inside grains a high density of dislocations was observed (Fig. 3a), and no changes in the volume fraction of the eutectic phase was detected (Fig. 3b).

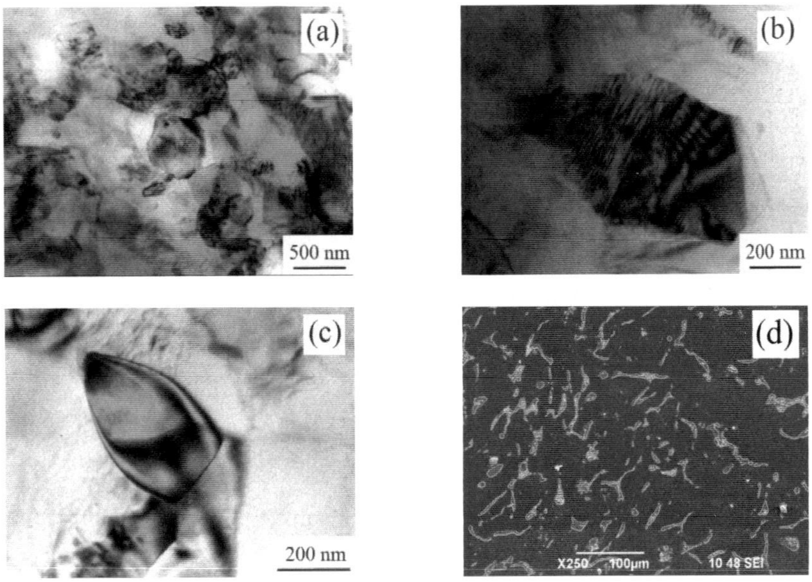

Figure 2. Microstructure of the Mg-2%Sr alloy after HPT: (a-c) TEM, (d) SEM

Studies of microstructure after HPT and additional annealing at a temperature of 200 °C revealed a strong decrease in the dislocation density (Fig. 4a). In the structure, fine grains were found with a size on the order of 0.8 ± 0.2 μm (Fig. 4a) and large ones with a size of 4 ± 1 μm (Fig. 4b). According to [11] the creation of such a bimodal structure should lead to both the high strength due to small grains and improved ductility at the expense of dislocation sliding in large ones. Changes in the volume fraction of eutectic component in this sample were not observed.

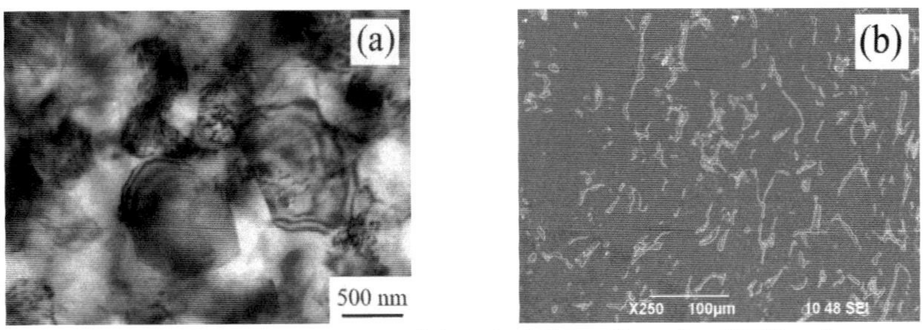

Figure 3. Microstructure of the Mg-2%Sr alloy after HPT and additional annealing at 150°C:(a) TEM, (b) SEM

Further development of recrystallization with the formation of a bimodal structure has been observed after additional annealing of the HPT samples at the temperatures of 250 (Fig. 5a) and 300°C (Fig. 6b). The eutectic with increasing temperature of additional annealing is transformed into single elongated conglomerates (Fig. 5b and 6b).

IOP Conf. Series: Materials Science and Engineering **380** (2018) 012014 doi:10.1088/1757-899X/380/1/012014

The microhardness of the initial samples was 42 ± 3 HV and increased after HPT up to 65 ± 5 HV, which is 1.5 times higher due to strong grain refinement. The study of thermal stability of the HPT samples by the microhardness measurements made it possible to establish that the samples are stable only up to temperature of 150°C (Fig. 7a). Thus, after additional annealing at 150°C, the microhardness remained practically unchanged, and after 200 °C a significant decrease to 53 ± 4 HV was found out. Further, a monotonic decrease in microhardness was observed.

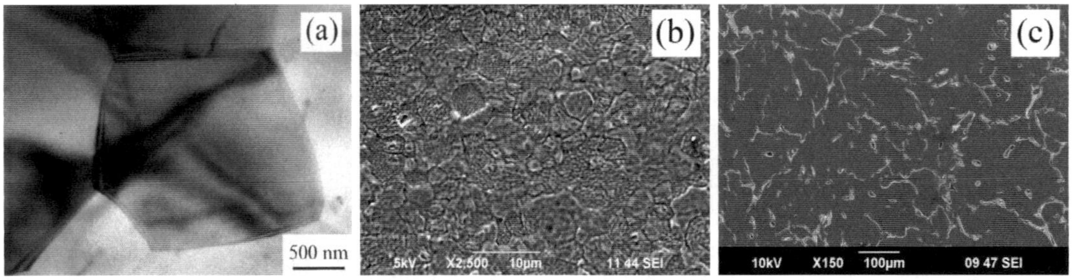

Figure 4. Microstructure of the Mg-2%Sr alloy after HPT and additional annealing at 200°C: (a) TEM, (b,c) SEM

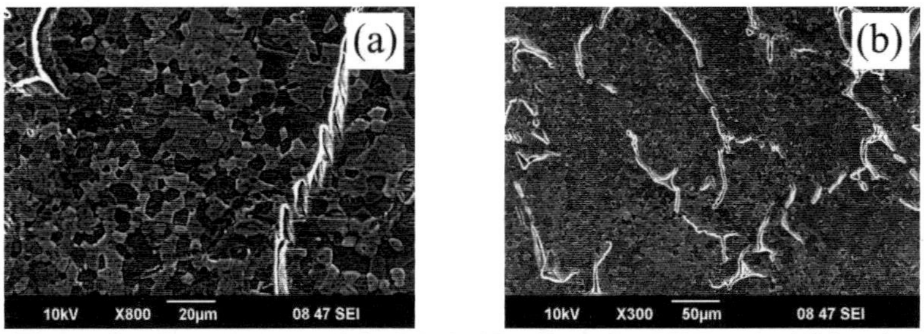

Figure 5. Microstructure of the Mg-2%Sr alloy after HPT and additional annealing at 250°C: (a, b) SEM

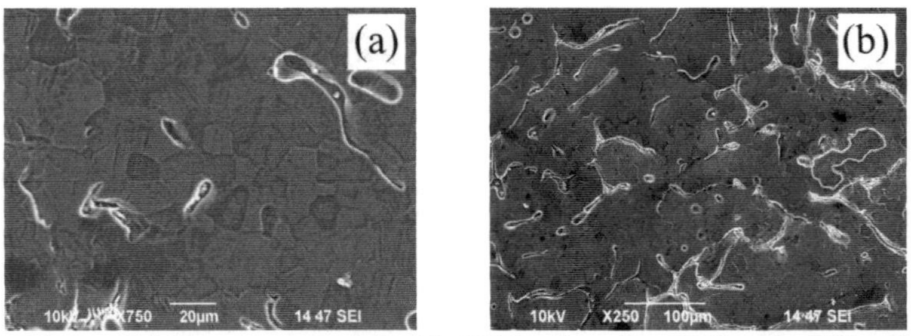

Figure 6. Microstructure of the Mg-2%Sr alloy after HPT and additional annealing at 300°C: (a, b) SEM

Figure 7b shows the results of tensile tests for the Mg-2%Sr alloy after various treatments. The initial samples after solid solution treatment do not have attractive mechanical properties compared to other magnesium alloys that have recently been developed for medical applications [8,9]. After the HPT and additional annealing at a temperature of 150°C, the samples have high microhardness, which implies the enhanced ultimate tensile strength. However, because of high internal stresses these samples showed brittle failure, not even reaching the yield stress. After additional annealing at the temperatures of 200, 250 and 300°C, the samples showed the enhanced strength and simultaneously residual ductility due to formation of a bimodal structure.

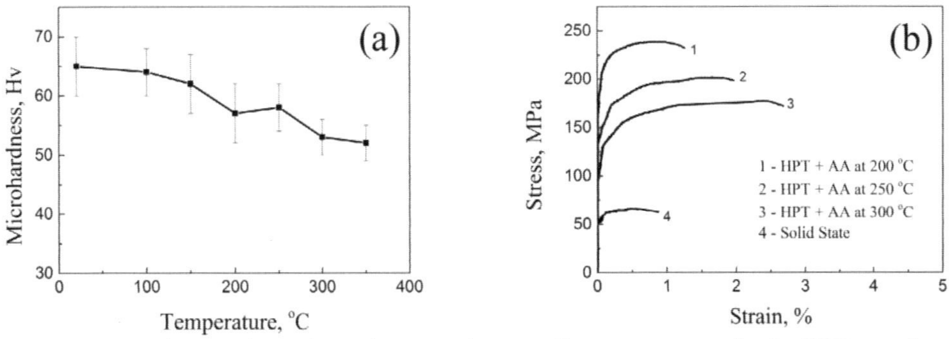

Figure 7. (a) – Microhardness dependence on the annealing temperature for the HPT samples of the Mg-2Sr alloy; (b) – results of tensile tests for the Mg-2Sr alloy at T=20°C and a strain rate of 10^{-3} s^{-1}

Studies of the corrosion resistance (Fig. 8a) of the Mg-2Sr alloy were carried out using samples subjected to HPT and additional annealing at 200°C in comparison with samples after solid solution treatment. It was found that the rate of the mass loss after solid solution treatment is higher than after the HPT and additional annealing. In particular the sample after solid solution treatment lost 72% of the initial mass after tests for 31 days, and the sample subjected to HPT and additional annealing was only 2%. Appearance of samples before the studies (0 days) and after exposure to saline (for days 15 and 27) are shown in Figure 8b. A decrease in the corrosion rate with grain refinement is typical for magnesium alloys [12-14]. In this study, an extra reason of less corrosion rate in ultrafine grained sample can be destroying at the HPT processing of a continuous grid of eutectics playing the role of a cathode at corrosion tests. So, the combination of enhanced strength and corrosion resistance observed in ultrafine grained Mg-2%Sr alloy can be attractive for next investigations of functional properties of the alloy to improve the design of implants in medicine.

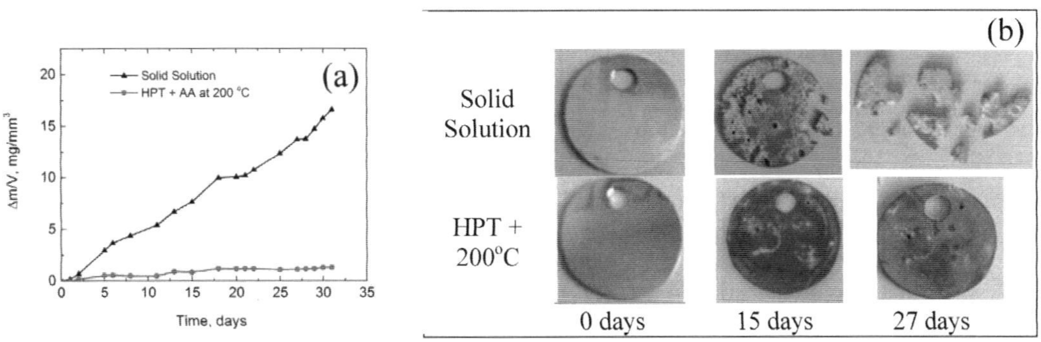

Figure 8. (a) The mass loss dependence on the exposure to saline of the Mg-2%Sr alloy; (b) appearance of the samples before (0 days) and after corrosion tests (15 and 27 days)

4. Conclusion

It was demonstrated that a combination of high pressure torsion and additional annealing at a temperature of 200°C leads to the formation of a bimodal structure providing the enhanced ultimate tensile strength of 245 MPa and residual ductility of 1.5%. In addition the grain refinement and destroying of a continuous grid of eutectics at the HPT processing allowed reducing the corrosion rate from 78 to 2% in 0.9% NaCl for 31 days. The significant increase in both the strength and corrosion resistance of ultrafine grained Mg-2%Sr alloy can be attractive for next investigations of functional properties of the alloy to improve the design of implants in medicine. Наши будущие исследования будут нацелены на выявление условий деформационно-термической обработки сплава которые обеспечат упрочнение не только за счет повышения плотности границ, но и за счет твердорастворного/дисперсионного упрочнения.

Acknowledgments

This work was supported by Russian Foundation of Basic Research (Project No. 16-53-53039 in cooperation with NSFC, China) and by the Ministry of Education and Science of the Russian Federation through a platform grant in the field of basic research (No. 16.7268.2017) and St. Petersburg grant 6.37.204.2016.

References

[1] Staiger M P, Pietak A M, Huadmai J and Dias G 2006 *Biomaterials* **27** 1728
[2] Li L, Zhang M, Li Y, Zhao J, Qin L and Lai Y 2017 *Regenerative Biomaterials* **129**
[3] Zheng Y Magnesium Alloys as Degradable Biomaterials 2016
[4] Pogorielov M, Husak E, Solodivnik A and Zhdanov S 2017 *Interv Med Appl Sci.* **9** (1) 27
[5] Bornapour M, Muja N, Shum-Tim D et al. 2013 *Acta Biomaterialia* **9** 5319
[6] Guo X N, Xie X N, Li N, Zheng Y F, Qin L 2012 *Acta Biomaterialia* **8** 2360
[7] Mingler B, Kulyasova O B et al. 2007 *Journal of Materials Science* **42** (5) 1477
[8] Kulyasova O B, Islamgaliev R K, Zhao Y and Valiev R Z 2015 *Adv. Eng. Mat.* **17** (12)
[9] Dobatkin S V, Lukyanova E A et al. 2017 *IOP Conf.Ser.: Mater. Sci. Eng.* **194** 012004
[10] Valiev R Z, Islamgaliev R K and Alexandrov I V 2000 *Progress Materials Science* **45** (2) 103
[11] Ma E 2006 *JOM* **58** (4) 49
[12] Argadea G R, PanigrahiaS K, Mishra R S 2012 *Corrosion Science* **58** 145
[13] Liu Y, Liu D, You C, Chen M 2015 *Frontiers of Materials Science* **9** (3) 247
[14] Saha P, Roy M, Datta M K, Lee B, Kumta P N 2015 *Mater Sci Eng C, Mater Biol Appl* **57** 294

IOP Conf. Series: Materials Science and Engineering **380** (2018) 012015 doi:10.1088/1757-899X/380/1/012015

Characterization of iron ore with Al_2O_3 as oxygen carrier in Chemical Looping Combustion (CLC)

N F Afandi[1], A Manap[1], S Mahalingam[2], and A Danial[1]

[1]Department of Mechanical, College of Engineering University Tenaga Nasional, Jalan IKRAM-UNITEN 43000, Kajang Selangor Malaysia
[2]Institute of Sustainable Energy (ISE), University Tenaga Nasional, Jalan IKRAM-UNITEN 43000, Kajang, Selangor, Malaysia

Abstract. Chemical looping combustion (CLC) is one of the carbon capturing technologies, in which oxygen from oxygen carrier reacts with fuel inside fuel reactor to produce CO_2 and H_2O. Fe-based oxygen carrier is widely used in CLC due to the low cost and less susceptible to carbon formation. This research focuses on synthesizing and characterizing iron ore with alumina in order to analyse the suitability of Malaysia iron ore as an oxygen carrier for CLC application. Iron ore with alumina was prepared using ball milling at various milling time which are 1 hour, 6 hours, 8 hours and 10 hours. The phase transformation, morphology and elemental composition of obtained samples were characterized using X-Ray Diffraction (XRD), scanning electron microscopy (SEM), and Energy Dispersive X-Ray (XRD), respectively. In CLC application, high reactivity of CLC can be obtained with the small particle size of oxygen carrier. This research succeeded in producing Fe_2O_3/ Al_2O_3 using ball milling with particle size that less than 10μm with crystallite size 17nm at 10 hours, which favourable to be used as oxygen carrier.

1. Introduction

Greenhouse gases that released to the environment cause the global warming and subsequent rise in atmospheric temperature. Many methods were used to control the emissions of carbon dioxide (CO_2) such as increase the efficiency of energy conversion, to use renewable energy and to capture the emissions of CO_2. However, capturing and storing the CO_2 (CCS) is the best method to reduce CO_2 emissions since combustion of fossil fuels will release the CO_2 into the air [1].

Chemical looping combustion was introduced as one of the carbon capture Storage (CCS) method that required low energy consumption and capable to increase efficiency of power plant comparing to conventional method [2]. CLC consists of two reactors, which are fuel reactor and air reactor. In CLC application, oxygen carriers play a very significant role in the overall process since oxygen carrier provides an oxygen for the combustion process. The oxygen carrier will react with the fuel in the fuel reactor in order to produce pure CO_2 and H_2O as a product. Then, this pure CO_2 can be stored in a liquid form instead release to the air. The reduced oxygen carrier will be flow into air reactor to oxidise with the air prior flow into fuel reactor again. Therefore, efficiency of CLC greatly depends on the performance of oxygen carriers. A good oxygen carrier should have good fluidization properties, high oxygen carrying capacity, high melting points, low cost, and environmental friendly [3].

The common metal based that was used as oxygen carriers are Fe-based, Cu-based, Ni-based, Mn-based and Co-based. From these five metals based, Fe-based, Cu-based and Mn-based are widely been applied in CLC system [4]. However, Cu-based has low melting point, meanwhile Ni-based shows a thermodynamic limitation, costly, and needed special safety measure to handle this material due to its hazardousness [5]. Fe-based is low cost, less susceptible to carbon formation, non hazardous, so the total operational cost would be low in CLC. However, particle agglomeration becomes major concern after several redox reactions when using Fe-based oxygen carriers [6]. Therefore, supported materials such as Al_2O_3, $MgAl_2O_4$, SiO_2, TiO_2 and YSZ were used with Fe-based in order to increase the durability of Fe-based oxygen carriers [7]. Iron ore was used continuously as Fe-based oxygen carrier in CLC since it shows good fluidization properties and minimum attrition [8].

Many methods were used to synthesize oxygen carrier such as mechanical mixing, dissolution method, freeze granulation method and sol-gel method [9-11]. However, mechanical mixing is a facile method that can produce small particle size of oxygen carrier. Song et al. [12] stated that, oxygen carrier with smaller particle size are more stable at high temperature and have superior mechanical properties. Hence, it can increase the efficiency of CLC.

Therefore, this research focuses on synthesizing and characterizing iron ore with Al_2O_3 using ball milling method and investigates the suitability of obtained iron ore from iron mining site in Malaysia as an oxygen carrier. In this research various ball milling time were varied in order to determine formation of Fe_2O_3 with Al_2O_3 since no extensive studies were found on synthesizing Malaysia iron ore with Al_2O_3 as an oxygen carrier for CLC application.

2. Methodology

In this research, ball milling method was used to produce Fe_2O_3/Al_2O_3 powder. Iron ore that found in Malaysia was heated up at temperature 950°C prior to the milling process. This process is to maximise the oxygen content in the powder that capable to carry more oxygen in CLC application[8]. The mixture of iron ore and alumina are inserted into the milling jar and milled with the speed of 300 rpm at various milled time which are 1 hour, 6 hours, 8 hours and 10 hours. Then, the obtained powder was characterized. The morphology and elemental composition of obtained samples were characterized using scanning electron microscopy (SEM), and Energy Dispersive X-Ray (XRD), respectively.

The crystalline phase of Fe_2O_3/Al_2O_3 powder was characterized using X-Ray diffraction (SHIMADZU 6000) at a scan speed 3°/min using CuKα radiation. The crystallite size was calculated by using Scherer equation. The Scherer equation is given below [13],

$$D = z\frac{0.9\lambda}{\beta \cos\theta} \tag{1}$$

where D is crystallite size (in nm), λ is the radiation wavelength (for CuKα radiation, λ 1.5418Å), θ is the diffraction peak angle and β is the broadening of the line ("half width") measured at half its maximum intensity (in radians). The morphology and elements of the Fe_2O_3/Al_2O_3 powder were observed using Scanning Electron Microscopy, SEM and Energy dispersive X-Ray, EDX respectively (JEOL).

3. Powder characterizations

3.1. Phase of obtained powder

The phase of obtained powder was analyzed using XRD analysis. Figure 1 shows XRD results of Fe_2O_3/Al_2O_3 when milled at 1 hour, 6 hours, 8 hours and 10 hours. The presence of Fe_2O_3 and Al_2O_3 can be observed in figure 1 corresponding to JCPDS no 39-1346 and JCPDS no 01-070-3319, respectively. However, the XRD reflection intensity decreases with increasing of milling time. The reduction of this intensity indicates the formation of amorphous structure that caused by the plastic deformation and disintegration of Fe_2O_3 and Al_2O_3. The XRD patterns shows only Fe_2O_3 and Al_2O_3 reflections, indicating that both compositions did not undergo significant reaction and physico-chemical changes [14]. Phourghahramani. P et al. stated that the X-Ray amorphization increases with increasing of milling time even with the particle size reduction and high BET surface area [14].

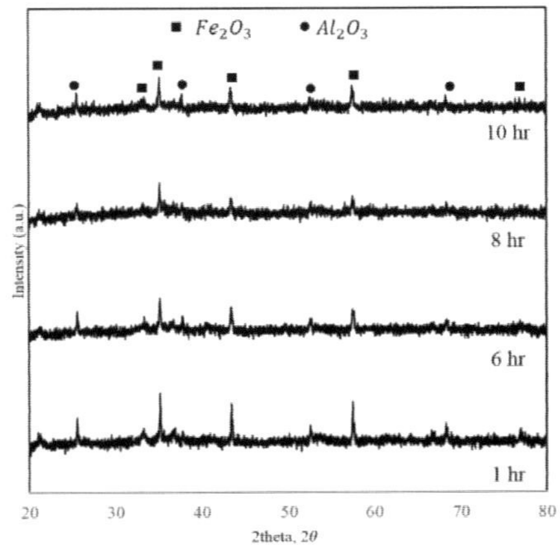

Figure 1. XRD result of Fe2O3 /Al2O3 when milling at 1 hour, 6 hours, 8 hours and 10 hours.

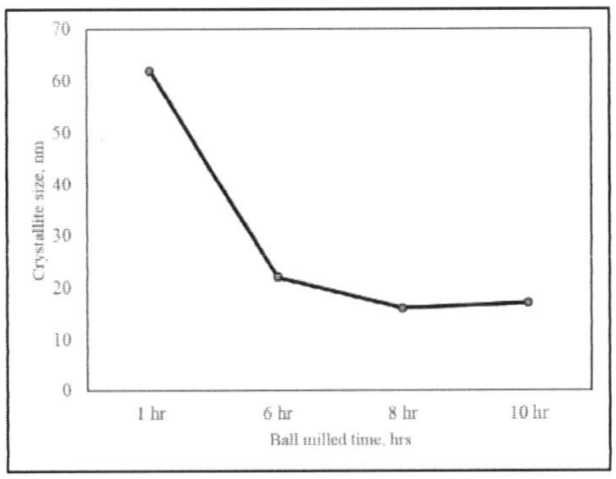

Figure 2. Average crystallite size of Fe2O3 /Al2O3 when milling at 1 hour, 6 hours, 8 hours and 10 hours.

3.2. Crystallite size

Figure 2 shows the average crystallite size of Fe_2O_3/ Al_2O_3 powder when milling at various milling time. The average crystallite size starts to decrease from 62 nm to 17 nm when milling time increases from 1 hour to 10 hours. The decrease in crystallite size is resulted from broken crystals even at the nanoscales due to the impact and abrasion forces generated inside milling jar [14].

3.3. Morphology of the Fe_2O_3/Al_2O_3

Figure 3 shows microstructure of Fe_2O_3/ Al_2O_3 when milled at 1, 6, 8 and 10 hours. It can be observed that particle size at 1 hour exhibit the aggregation of large irregular particle. Particle size start to decrease when milling time was increased from 1hour to 10hours. This is due to the effect of impact force and shear friction that generate cracks continuously and decrease the particle into small particle size [15]. However, no observable effect on the particle morphology when increases the milling time from 1hour to 10hours. Therefore, increases the milling time capable to produce Fe_2O_3 and Al_2O_3 at small particle size, hence high BET surface area. Song et al. [12] stated that, smaller particle of oxygen carriers are more stable at high temperature, and have superior mechanical properties which can increase CLC performance.

IOP Conf. Series: Materials Science and Engineering **380** (2018) 012015 doi:10.1088/1757-899X/380/1/012015

Figure 3. SEM images of Fe_2O_3 /Al_2O_3 when milling at different time a) 1hr b) 6hrs c) 8hrs d) 10hrs.

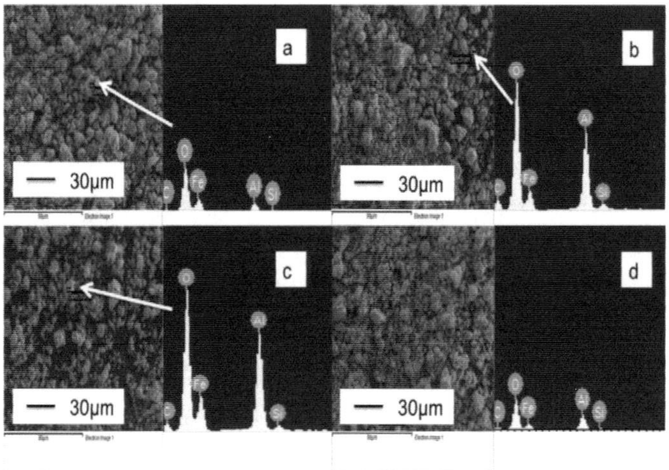

Figure 4. Elemental composition of Fe_2O_3 /Al_2O_3 when milling at different time a) 1hr b) 6hrs c) 8hrs d) 10hrs.

3.4. Elements of the Fe_2O_3/Al_2O_3

Figure 4 shows the elemental composition of Fe_2O_3/ Al_2O_3 when milling at different time. Increases milling time does not affect the elemental composition of the powder. The presence of Fe, Al and O can be seen at all these milling times. In addition, the presence of Si and C can be found in these powders which is in agreement with Wong et al. [16]. Si would not affect the performance of CLC [17]. Therefore, it can be negligible. Presence of C was due to the contamination from milling process and carbon tape during sample preparation prior analysis. Hence, this Fe_2O_3 powder is fit to be tested in thermogravimetric analysis (TGA) for simulating the CLC application.

4. Conclusions

According to the obtained result, Fe_2O_3 and Al_2O_3 was successfully synthesized using iron ore with alumina by ball milling mixing method. Fe_2O_3/Al_2O_3 that was milled at 10hrs exhibit smaller particle size which is less than 10μm with crystallite size 17nm compare to the other milling time. Smaller particle size can increase CLC performance, hence produced oxygen carrier that can be used in CLC application. Furthermore, this Fe_2O_3 with Al_2O_3 can be tested using TGA analysis in order to simulate the CLC process for its performance.

5. Acknowledgments
The authors acknowledge the financial supports by the UNITEN (Grant No. RJO10289176/B/1/2017/9) and Malaysian Ministry of Higher Education (Grant No. FRGS20160105).

6. References
[1] Yamasaki A 2003 An overview of CO2 mitigation options for global warming- emphasizing CO2 sequestration options *J. Chem Eng of Japan* **36** 361-375
[2] Shijaz H, Attada Y, Patnaikuni VS, Vooradi R, Sarath BA 2017 Analysis of integrated gasification combined cycle power plant incorporating chemical looping combustion for environmental friendly utilization of Indian coal *Eneryg Converse Manag* **151** 414-425
[3] Adánez J, Abad A, Garc a-Labiano F, Gayán P, de Diego LF 2012 Progress in Chemical looping combustion and reforming technologies. A review *Prog Energy Combust Sci* **38** 215-282
[4] Wang B, Yan R, Lee DH, Zheng Y, Zhao H, Zheng C 2011 Characterization and evaluation of Fe_2O_3/Al_2O_3 oxygen carrier prepared by sol-gel combustion synthesis *J Anal Appl Pyrolysis* **91** 105-113
[5] Tang M, Xu L, Fan M 2015 Progress in oxygen carrier development of methane based chemical looping reforming. A review *Applied energy* **151** 143-156
[6] Nandy A, Loha C, Sai G, Sarkar P, Karmakar K 2016 Present status and overview of Chemical Looping Combustion technology *Renew Sust Energ Rev* **59** 597-619
[7] Gao ZF, Wu ZJ, Liu WM 2016 Preparation and chemical looping combustion properties of Fe_2O_3/ Al_2O_3 derived from metallurgy iron-bearing dust. *J Environ Chem Eng* **4** 1653-1663
[8] Haider SK, Azimi G, Duan L, Anthony EJ, Patchigolla, Oakey JE, Leion H, Mattisson T, Lyngfelt A 2016 Enhancing properties of iron ore and manganese ores as oxygen carriers for chemical looping combustion *Appl Energ* **163** 41-50
[9] Zhao HB et al. 2008 $NiO/NiAl_2O_4$ oxygen carrier prepared by sol-gel for chemical looping combustion fuelled by gas *J Fuel Chem Tech* **36** 261-266
[10] Ishida M, Yamamoto M and Obata T 2002 Experimental results of chemical looping combustion with $NiO/NiAl_2O_4$ particle simulation at 1200°C *Energ Converse Manag* **43** 1469-1478
[11] Cho P, Mattinson T, Lyngfelt 2004 Comparison of iron-, nickel-, copper- and manganese based oxygen carriers for chemical looping combustion *Fuel* **83** 1215-1225
[12] Song Q, Xiao R, Deng Z, Zhang H, Shen L, Xiao J 2008 Chemical looping combustion of methane with $CaSO_4$ oxygen carrier in a fixed bed reactor *Energ Converse Manag* **49** 3178-3187
[13] Parra MR and Haque FZ 2014 Aqueous chemical route synthesis and the effect of calcination temperature on the structural and optical properties of ZnO nanoparticles *Journal of Materials Research and Technology* **3** 363-369.
[14] Pourghahramani P, Forssberg E 2006 Comparative study of microstructural characteristic and stored energy of mechanically activated hematite in different grinding environments *Int. J. Miner. Process* **79** 130-139
[15] Wang B, Wei S, Wang Y, Guo L, Xue J, Pan, F, Tang A, Xu B 2018 Effect of milling time on microstructure and properties of Nano-titanium polymer by high energy ball milling *Appl Surf Sci* **434** 1248-1256
[16] Wong YW, Woon HS, Tan CY 2014 Purification and conversion of Malaysian iron ores into industrial grade iron oxide colour pigment *Mater Res Innov* **18** 159-163, 2014.
[17] Linderholm C, Schmitz S, 2016 Chemical-looping combustion of solid fuels in a 100kW dual circulating fluidized bed system using iron ore as oxygen carrier *J Environ Chem Eng* **4** 1029-1039

IOP Conf. Series: Materials Science and Engineering **380** (2018) 012016 doi:10.1088/1757-899X/380/1/012016

Hydrogen damage in metals

L Gu and X Y Li

Department of Mechanical Engineering and Applied Electronics Technology, Beijing University of Technology, Beijing, 100124, P.R. China

Abstract. In corrosive environment, the hardness along the depth direction and crack propagation direction of specimen is affected by hydrogen concentration and stress. In this paper, micro-hardness tester was employed to quantitatively evaluate the hydrogen distribution in metals. Hydrogen saturated value and hydrogen saturated layer of specimens were obtained. The experimental results demonstrated that hydrogen concentration behavior in a known equibiaxial stress environment can be verified experimentally using micro-hardness technique. There is a simple additive relationship between the hydrogen-induced micro-hardness increment and the stress-induced micro-hardness increment in Vickers micro-hardness measurement. The hydrogen distribution of specimens was analyzed by taking the change of the micro-hardness increment along the depth direction and crack propagation direction of specimens as the indicator.

1. Introduction

Metal corrosion damage can cause catastrophic accidents, resulting in huge economic losses. Hydrogen is one of main corrosive medium in the metal industry. It is expected that hydrogen will be used as an energy carrier in the near future since it is clean, movable and abundant, however, hydrogen always has a deleterious effect on their mechanical properties, e.g. hydrogen-assisted stress corrosion cracking[1] and hydrogen embrittlement[2]. The effect of hydrogen on the metallic material begins when the hydrogen enters the metal. Once the hydrogen enters the metal, it can diffuse through the metal or get entrapped at microstructural features and hydrogen-trapping sites. Changes in the concentration distribution of the diffusible hydrogen are likely to cause changes in the mechanical properties of metals[3,4]. The crux of hydrogen damage study is to accurately evaluate the degree of hydrogen distribution within metals. It is therefore important to analyze the time and applied stress variation of hydrogen distribution in metals to understand the effects of hydrogen on the mechanical properties of materials. However, experimental evaluation of the non-homogeneous hydrogen concentration distribution behavior in a local area such as the vicinity of a crack is very difficult, due to the light weight and high mobility of hydrogen. Thermal desorption analysis can only evaluate the total hydrogen content and separate hydrogen trapping sites. The hydrogen microprint technique can visualize the presence of hydrogen in a local area[5], and Nagao used this technique to observe the effect of the stress fields caused by a bending test on hydrogen transport[6]. However, the hydrogen microprint technique cannot quantitatively evaluate the hydrogen content although it is no doubt a necessary technique for investigating hydrogen trapping and transport in local areas. Thus, an unconventional and quantitative method focusing on a local area, e.g., the vicinity of a crack tip, needs to be established. Until now, the local hydrogen concentration behavior has been evaluated by numerical simulation according to the diffusion equation considering time or stress fields. There are many publications that report numerical simulations of the hydrogen concentration behavior in the metal[7]. However, this is just an assumption and no experimental research has confirmed this. In order to determine the mechanism for hydrogen damage in metals, the mechanism by which hydrogen distribution in the local area has to be clarified experimentally.

The present study is divided into two parts. First, we use layer stripping micro-hardness method to analyze the hydrogen distribution along the depth direction of the hydrogen-charged specimens without applied stress under different corrosion periods, then to analyze the hydrogen distribution under different applied stress fields with the same corrosion period. In addition, the theoretical time and applied stress variation of the hydrogen distribution in the hydrogen-charged specimen is calculated using models on the basis of Fick's law. The time and applied stress variations of the experimental values and of the theoretical values are compared.

2. Experimental Methods

2.1. Experimental design

First, 12 cylindrical specimens with the same diameter of 18 mm and height of 12 mm were soaked in saturated hydrogen sulfide solution for four different corrosion periods (24 h, 48 h, 72 h, 96 h). The experiment was divided into 4 groups, each group with three specimens selected for each test to verify the reproducibility of the material. Ten hardness points were measured to take the average for each test. Starting from the surface, the micro-hardness values were obtained at regular intervals of 50 μm progressing into the thickness of the specimen. Hardness was measured until it reached a constant value. Tested the original hardness of the specimen after grinding the specimen surface before corrosion. The hardness values of each test plane and the depth values of the stripping layer corresponding to the plane were recorded using the layer stripping micro-hardness method after corrosion.

Second, 6 Modified wedge-opening-loading (WOL) specimens were used in this study. Parameters of the WOL specimen size is shown in Figure 1. The constant stress intensity factor at the crack tips was fixed by the bolt. Then we could get the non-uniform applied stress field along the crack propagation direction of the specimens. Starting from the crack tip, the micro-hardness values were obtained at intervals of 80~100 μm progressing along the crack propagation direction of the specimen. Hardness was measured until it reached a constant value. 10 random hardness points of the specimen surface far from the crack tip were measured to determine the hardness value of the unstressed state. Then soaked the specimens in saturated hydrogen sulfide solution for 48 h and repeated the testing process on the specimen surface.

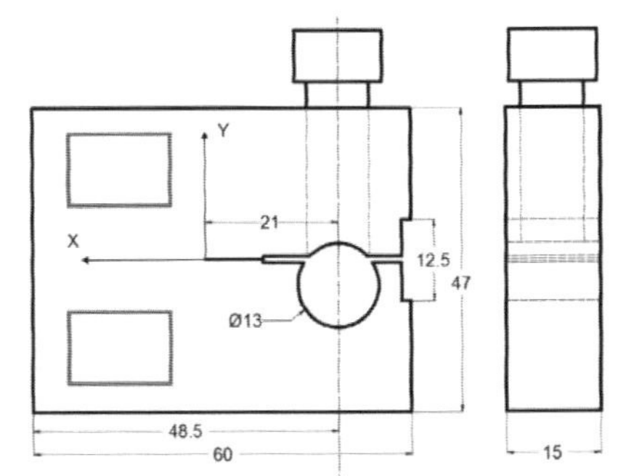

Figure 1. The Modified WOL specimen.

2.2. Test material and experimental apparatus

The test material was #45 steel, its measured tensile strength was 714 MPa, and the yield strength was 408 MPa. The chemical composition of the materials is shown in Table 1.

IOP Conf. Series: Materials Science and Engineering **380** (2018) 012016 doi:10.1088/1757-899X/380/1/012016

Table 1. Chemical compositions of #45 steel (mass %)

C	Si	Mn	S	P	Cr	Ni
0.46	0.27	0.65	0.015	0.018	0.20	0.214

Experimental apparatus was a universal hardness tester produced in Germany Zwick Company, its model was Zwick Roell ZHU2.5. The test method was according to the GB/T4340.1.
In this study, Vickers hardness HV0.2 was chosen as the hardness index of the material, which could meet the highly sensitive experimental requirements.

2.3. Test environment
H_2S gas was produced by the reaction of 30% Na_2S solution with 30% H_2SO_4 solution. Hydrogen sulfide was dissolved in the American Society of Corrosion Engineers NACE standard A standard solution (containing 5% sodium chloride and 0.5% glacial acetic acid). Specimens were soaked into the hydrogen sulfide corrosion environment after the solution concentration was stable. Used acetone to remove grease and rust from the surface of the specimens before soaking.

3. Experimental results
The distribution curves of the micro-hardness increment along the depth direction of specimens in each group after soaking in saturated hydrogen sulfide solution for various corrosion periods (24 h, 48 h, 72 h, 96 h) are shown in Figure 2. The curves under the same corrosion condition have a strong consistency, indicating the stability of the corrosion soaking in the hydrogen sulfide solution. Figure 3 also shows that the micro-hardness increment increases first and then decreases until micro-hardness increment returns to zero along the depth direction, indicating that there is a clear stratification phenomenon along the depth direction of specimens after soaking in hydrogen sulfide solution.

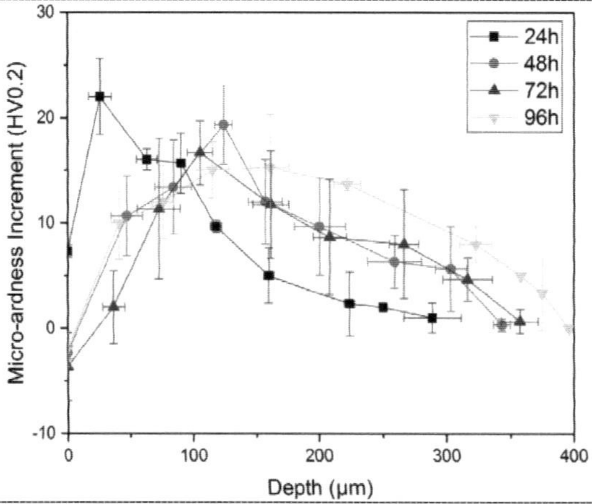

Figure 2. Comprehensive comparison curves.

The distribution curves of the micro-hardness increment along the crack propagation direction of specimens before and after soaking in saturated hydrogen sulfide solution are shown in Figure 3.

IOP Conf. Series: Materials Science and Engineering **380** (2018) 012016 doi:10.1088/1757-899X/380/1/012016

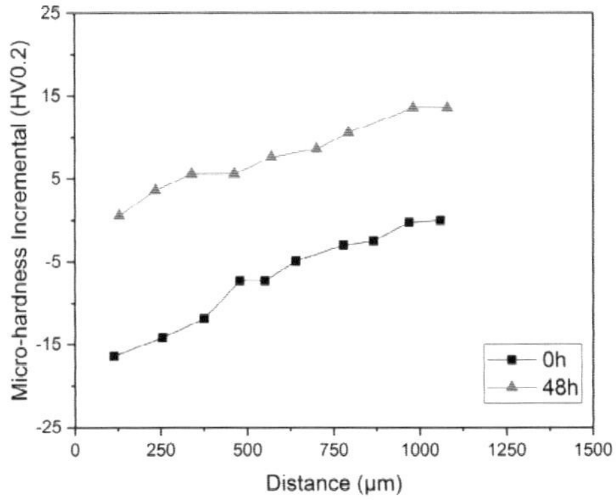

(a) Specimen 1.

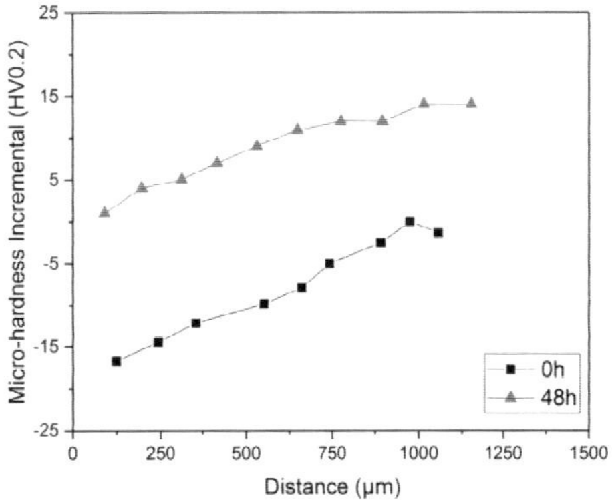

(b) Specimen 2.

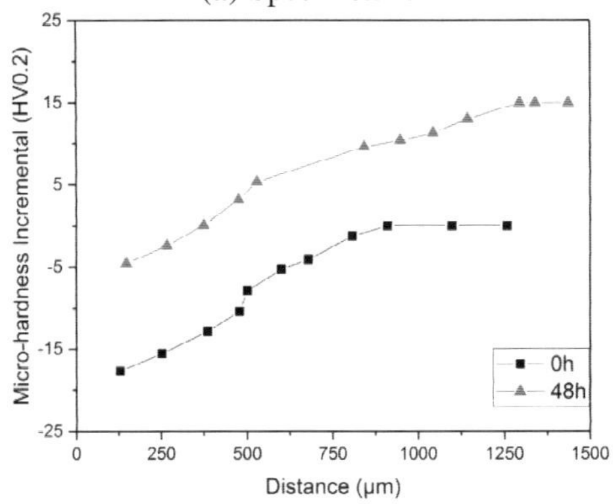

(c) Specimen 3.

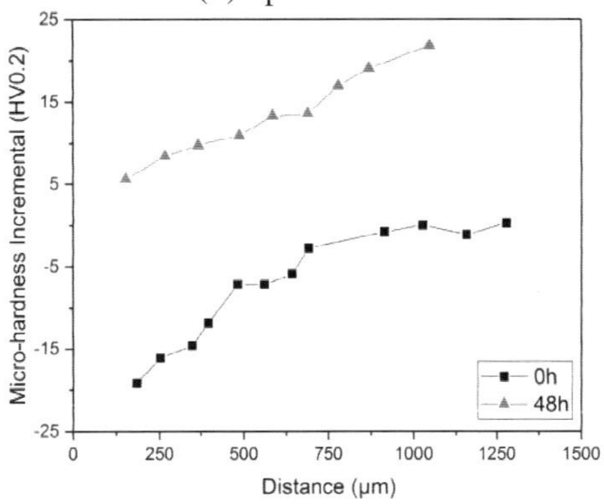

(d) Specimen 4.

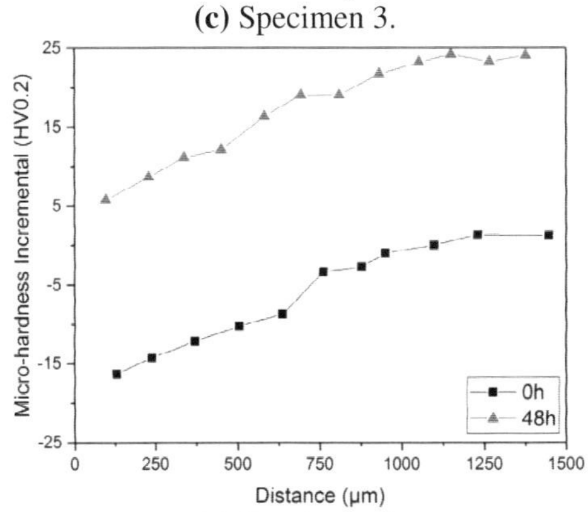

(e) Specimen 5.

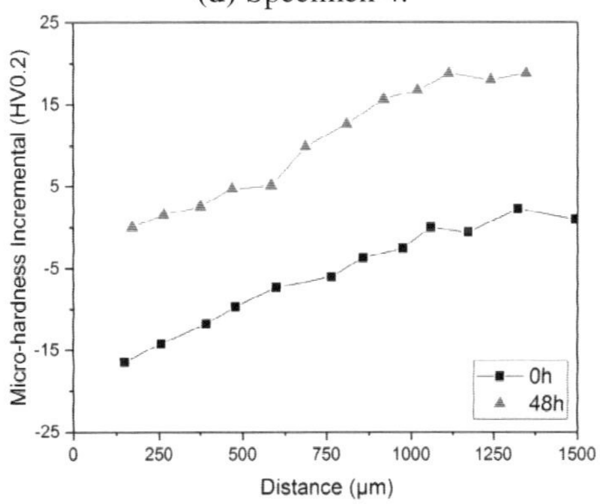

(f) Specimen 6.

Figure 3. Distribution curves of the micro-hardness increment.

4. Discussion

4.1. Hydrogen saturated value and hydrogen saturated layer

Under the condition of varies corrosion periods, the average values of maximum micro-hardness increment of the test specimens is almost the same, the mean is 17.94 HV0.2, and the standard deviation is 0.98. This shows that there are no significant difference in the maximum micro-hardness increment under varies corrosion periods.

Since the maximum micro-hardness increment of the metal under the condition of varies corrosion periods are almost constant, and due to that increase in hydrogen concentration is proportional to the increase in hardness[8], the hydrogen concentration corresponding with the micro-hardness increment can be defined to saturated state. That is, the hydrogen concentration reaches the saturated concentration (solubility) c_{max} in metal. For the same material, c_{max} is a constant. Because the present study focus on the analysis of hydrogen distribution in metals, the behavioral analysis of hydrogen in metal can be analyzed by the distribution of the micro-hardness increment instead of the hydrogen concentration.

Put the diffusion coefficient D, the average maximum hardness increment and the corrosion periods (24 h, 48 h, 72 h, 96 h) into the Eq. (1). The corresponding theoretical micro-hardness increment distribution curves of hydrogen diffusion in the metal are obtained. Compare the theoretical micro-hardness increment distribution curves with the experimental increment distribution curves according to the corrosion periods 24 h, 48 h, 72 h, 96 h, respectively, as shown in Figure 4.

$$\Delta HV(x,t) = \Delta HV_s\left[1 - erf\left(\frac{x}{2(Dt)^{1/2}}\right)\right] = \Delta HV_{max}\left[1 - erf\left(\frac{x}{2(Dt)^{1/2}}\right)\right] \tag{1}$$

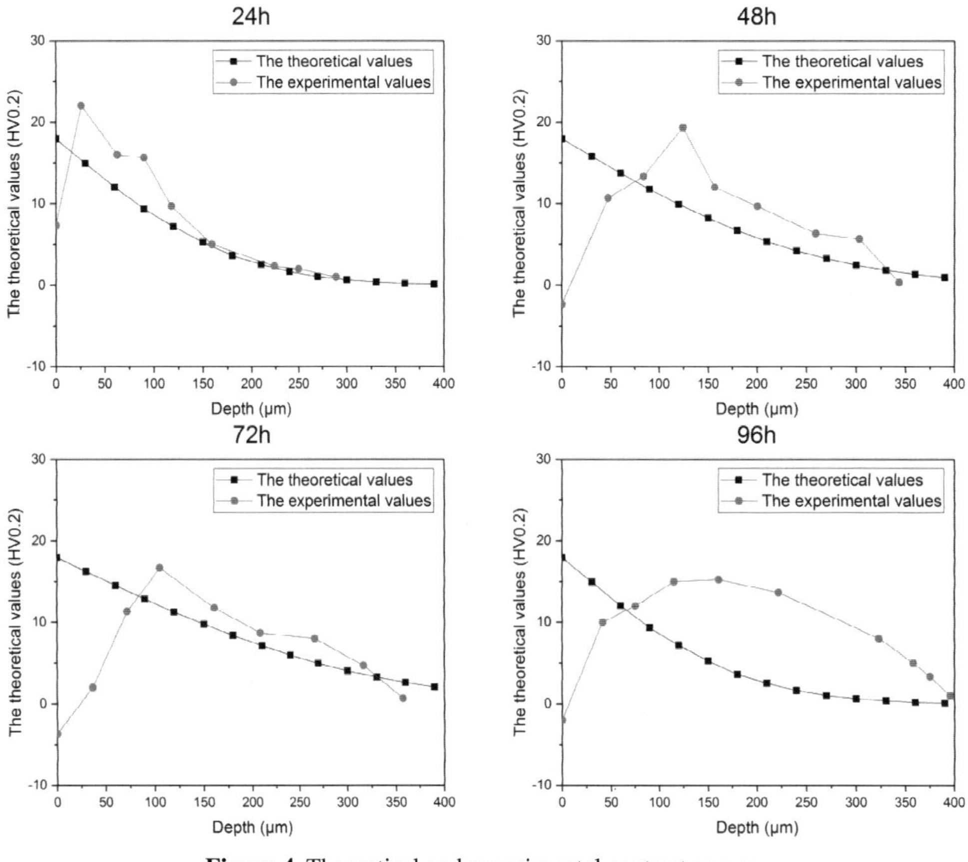

Figure 4. Theoretical and experimental contrast curves.

The present study used the hydrogen diffusion coefficient determined by J. P. Hirth [9] to calculate the macroscopic hydrogen distribution for materials.

Since the theoretical curves of hydrogen diffusion are analyzed using a numerical simulation according to Fick's second law. The curves are always from big to small along the depth direction using the hydrogen diffusion theory. This is inconsistent with the actual situation as shown in Figure 4.

It can be seen that the big-to-small section of experimental curves has a certain distance from the theoretical curves along the depth direction.

Keep the maximum value and move the theoretical curves along the depth direction until the theoretical curves coinciding with the experimental curves and keep the value of the front distance as the maximum micro-hardness increment. The moving distance for each corrosion period as follows: 40 μm to the right for 24 h, 70 μm to right for 48 h, 80 μm to right for 72 h, 85 μm to right for 96 h. Therefore, a group of new theoretical curves are obtained as shown in Figure 5.

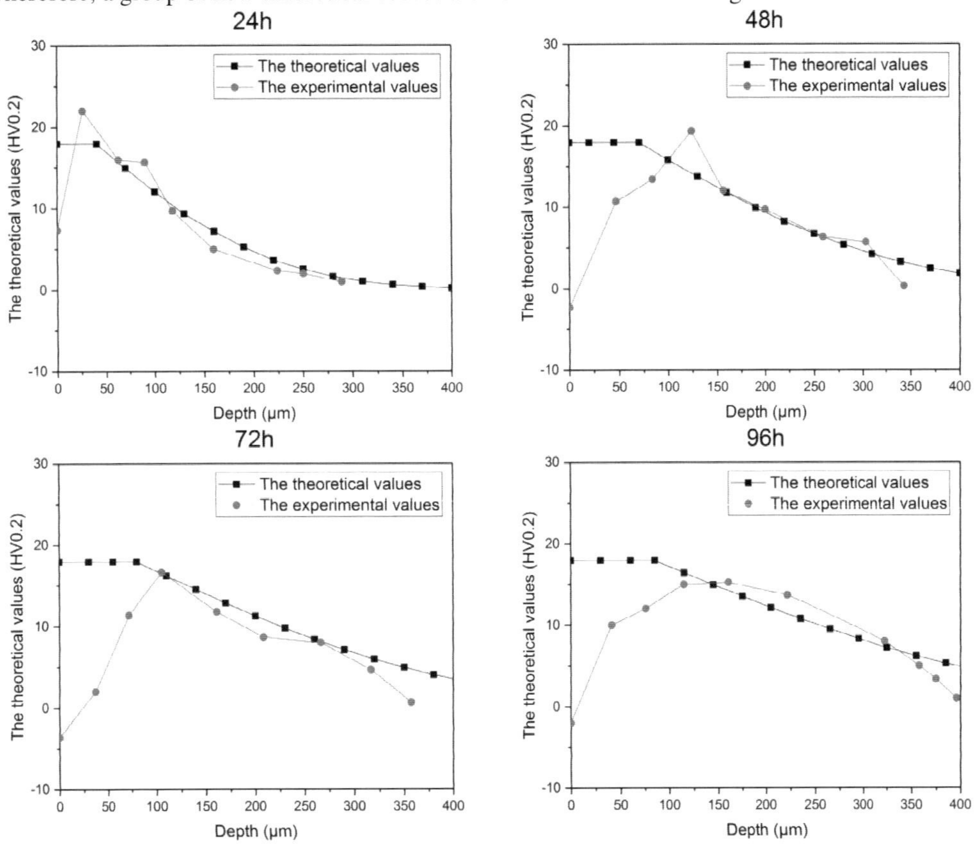

Figure 5. Theoretical and experimental contrast curves.

It can be clearly seen from Figure 5 that the theoretical curves after translation is in good agreement with the experimental curves, indicating that the hydrogen saturated layer is appeared along the depth direction of hydrogen-charged specimens. The presence of hydrogen saturated layer can be speculated that hydrogen can get entrapped and generate at the microstructural trap sites in the specimens[10]. The thickness of hydrogen saturated layer increases as the corrosion period increases.

It can be seen from figure 6 that the hydrogen concentration in the metal gradually decreases along the depth direction until zero after the hydrogen saturated layer. The trend of changes is consistent with the theoretical one for the same corrosion period.

4.2. Hydrogen distribution with applied stress

The distribution curves of the micro-hardness increment along the crack propagation direction of specimens before soaking in saturated hydrogen sulfide solution is shown in Figure 6.

In the non-corrosive environment, the micro-hardness increment versus the applied stress fields have strong linear correlation with a correlation coefficient, R, of 0.98 and significance level of 0.04×10^{-14}.

Based on the relationship between hardness and stress, we extracte the experimental hydrogen distribution from the applied stress fields of hydrogen-charged specimens. Then the applied stress variations of the experimental hydrogen-induced micro-hardness increment and of the theoretical value are compared. The trend of changes is consistent along the crack propagation direction of specimens, as is shown in Figure 7.

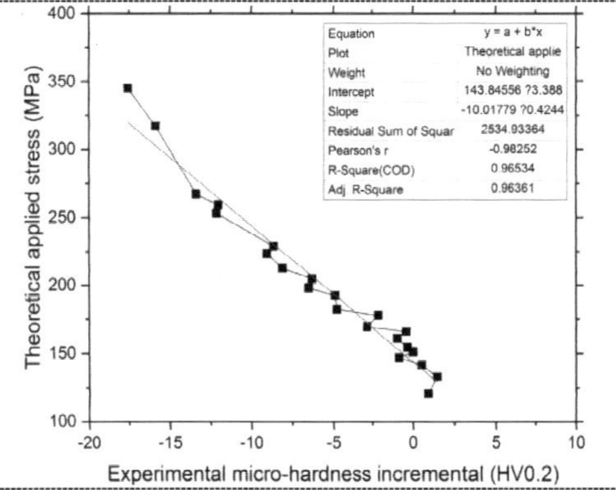

Figure 6. Linear relationship between the experimental micro-hardness increment and the theoretical applied stress fields.

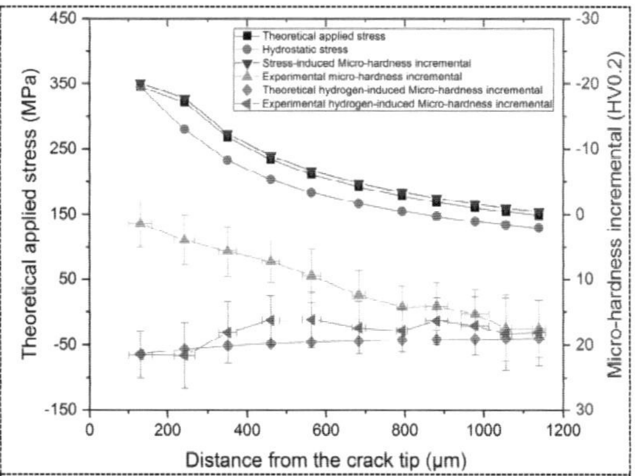

Figure 7. Theoretical and experimental contrast curves of micro-hardness increment – distance from the crack tip.

5. Conclusions

The micro-hardness tests for various corrosion periods or various applied stress fields were performed at room temperature. Took the change of the micro-hardness increment along the depth direction or the crack propagation direction of specimens as an index, then analyzed and evaluated the hydrogen distribution in the metal after soaking in hydrogen sulfide solution. Our conclusions are summarized as follows.

- There was a stable saturated hydrogen concentration (solubility) in the metal in hydrogen sulfide solution of various corrosion periods. And there is a hydrogen saturated layer hydrogen saturated layer along the depth direction of specimens pre-charged with hydrogen. The thickness of hydrogen saturated layer increases as the corrosion period increases in specimens.
- There is a simple additive relationship between stress-induced micro-hardness increment and hydrogen-induced micro-hardness increment in Vickers micro-hardness measurement.
- The hydrogen concentration behavior in a known equibiaxial stress environment can be verified experimentally by micro-hardness tester. References

[1] G L. Pioszak and R P. Gangloff. Hydrogen Environment Assisted Cracking of a Modern Ultra-High Strength Martensitic Stainless Steel [J]. CORROSION, 2017, 73(9): 1132–1156.

[2] P.A.S. Pereira, C.S.G. Franco, J.L.M. Guerra Filho and D.S. dos Santos. Hydrogen effects on the microstructure of a 2.25Cr- 1Mo-0.25 V steel welded joint [J]. *International Journal of Hydrogen Energy*, 2015, 40(47): 17136–17143.

[3] Y. Murakami, T. Kanezaki and Y. Mine. Hydrogen effect against hydrogen embrittlement. *Metall Mater Trans* A, 2010, (41A): 2548–2562.

[4] S. Wang, S. Ohnuki, N. Hashimoto and K. Chiba. Hydrogen effects on tensile property of pure iron with deformed surface. *Mater Sci Eng* A, 2013, (560): 332–338.

[5] J. O García. Hydrogen microprint technique in the study of hydrogen in steels. *J Mater Sci*, 1985, (20): 2623–2629.

[6] A. Nagao, S. Kuramoto, K. Ichitani and M. Kanno. Visualization of hydrogen transport in high strength steels affected by stress fields and hydrogen trapping. *Scr Mater*, 2001, (45): 1227–1232.

[7] O. Takakuwa, M. Nishikawa and H. Soyama. Numerical simulation of the effects of residual stress on the concentration of hydrogen around a crack tip. *Surf Coat Technol*, 2012, (206): 2892–2898
[8] P. Kumar and R. Balasubramaniam. Determination of hydrogen diffusivity in austenitic stainless steels by subscale microhardness profiling [J]. *Journal of Alloys and Compounds*, 1997, 255(1–2): 130–134.
[9] S.X. Xie and J.P. Hirth. Permeation of Hydrogen, Trapping, and Damage in Spheroidized AISI 1090 Steel [J]. CORROSION, 1982, 38(9): 486–493.
[10] K Esaklul. Hydrogen damage. In book: Trends in Oil and Gas Corrosion Research and Technologies, first ed., A.M. EI-Sherik, Eds.; Woodhead Publishing: United Kingdom, 2017; Chapter 13, pp.315–340.

Preliminary application of cerium-based hollow nanospheres as drug carrier

J C Li[1, 2, 3], H Jiang[1, 2, 3] and X M Wang[1, 2, 3]

[1]School of Biological Science and Medical Engineering, Southeast University, Nanjing, 210096, China
[2]State Key Lab of Bioelectronics (Chien-Shiung Wu Laboratory), Southeast University, No. 2 Sipailou, Nanjing, 210096, China
[3]National Demonstration Center for Experimental Biomedical Engineering Education, Southeast University, Nanjing, 210096, China

E-mail: xuewang@seu.edu.cn

Abstract. As more and more chemical activity and biological activity are found, cerium, one of the largest reserves of cheap rare earth lanthanides, is widely used in many fields such as flint, decolorizer, abrasive polish, glass additives, reducing agents, catalysts, harmful metal substitutes and other special functional materials. Cerium nanoparticles can be used in cardiovascular diseases, neurodegenerative diseases and radiation-induced tissue damage, also be used as biomimetic enzymes, biocatalysis, biomedicine, biological scaffolding and other biological fields. However, cerium-based hollow nanospheres have never been properly tested as a drug carrier. In this report, we explored the potential of cerium-based materials as a drug carrier which is a brand new attempt to expand the scope of application of cerium-based nanomaterials in the field of biomedicine.

1. Introduction

Due to its oxygen storage capacity via facile Ce^{4+}/Ce^{3+} redox cycles, high mechanical strength, and unique optical properties, the cerium elemental oxygen compound (ceria) as one of the most important functional rare earth oxides is a promising material widely be used in many fields such as catalysts[1], corrosion prevention[2], fuel cells[3, 4], oxygen sensors[5], magnetic materials[6], ultraviolet blocks[7], and optical materials[8]. In addition to these traditional applications, more and more studies have shown that cerium oxide compound has many biological activities and can be applied in the biomedical field. Ceria can mimic the activity of key antioxidant enzymes, for instance, superoxide dismutase (SOD) and catalase (CAT) mimetic activity. The mechanism potentially abates noxious intracellular reactive oxygen species (ROS) via the Ce^{4+}/Ce^{3+} self-regenerating redox cycles which might fight chronic inflammation and the pathologies associated with oxidative stress including cancer and neurodegeneration[9-13]. The status of cerium-based nanomaterials in the biomedicine field will be more and more important.

Like other inorganic hollow nanospheres, cerium-based hollow nanospheres also have the unique chemical and physical properties, which make them have obtained the most attention for medicinal applications[14,15]. The large void space in hollow structures have been successfully used for adsorption, enrichment, protection and release of sensitive materials and biologically active species[16-20]. So they can be developed into potential intelligent drug carriers which can maintain the concentration of drugs in the precise sites of the body, improve the therapeutic efficacy and reduce toxicity[21-23]. However, cerium-based hollow nanospheres have never been properly tested as a drug carrier.

Here in our study, cerium-based hollow nanospheres were synthesized with a typical method[24] and then have been tried to carry a model medicine, protoporphyrin, as a fluorescent indicator into the cell and reached very good repercussion. The carrier can be encapsulated by chitosan which to obtain pH response function for tumor targeting delivery in vivo and release from the acidic microenvironment around tumor. We explored the potential of cerium-based nanospheres as drug carrier for the first time which is a brand new attempt to expand the scope of application of cerium-based nanomaterials in the field of biomedicine.

2. Experimental section

The cerium hollow nanospheres were synthesized by the previously reported method[24] and characterized by electron microscopy. A rapid colorimetric assay for cellular growth and survival, the tetrazolium-based colorimetric assay (MTT), was applied to measure proliferation and cytotoxicity for evaluation of cytotoxicity in vitro. Erythrocyte protoporphyrin was used as an indicator drug for preliminary testing and evaluation of drug delivery ability of cerium hollow nanospheres by microscopy to observe fluorescence. The specific experimental procedure has shown in the schematic (Figure 1).

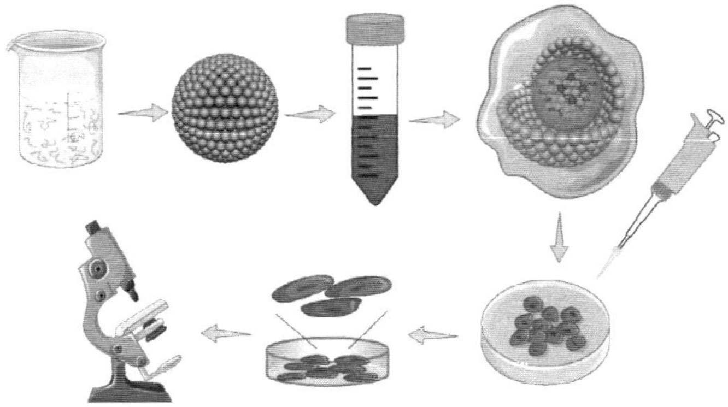

Figure 1. Schematic illustration of the experimental procedure

3. Results and discussion

3.1. Characterization

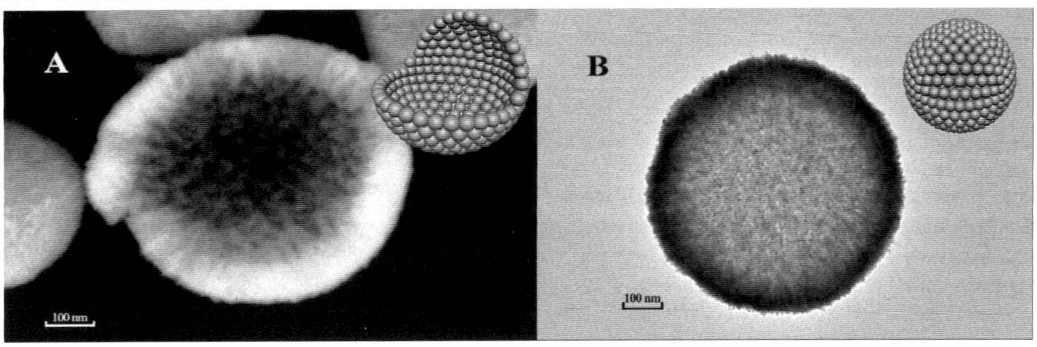

Figure 2. Characterization by electron microscopy: A. SEM image B. TEM image

The morphology and internal structure of the cerium hollow nanospheres were characterized by scanning electron microscopy (SEM) and transmission electron microscopy (TEM). The SEM and TEM images in figure 2 revealed that these hollow-structured nanospheres had the mesoporous shells that were composed of a large amount of uniform nanoparticles. Nanoshell structure fragments microphotograph in the figure 2A indicated that the shell of cerium hollow nanospheres were formed by the aggregation of many small spherical cerium nanoparticles and filled with a lot of mesoporous between them. The TEM microphotograph (Figure 2B) showed the obvious hollow structure inside cerium nanospheres. The disordered mesostructure of the surface and the internal hollow structure of make the nanospheres have excellent mass-storage properties for loading and adsorption capacity.

3.2. The safety evaluation in vitro

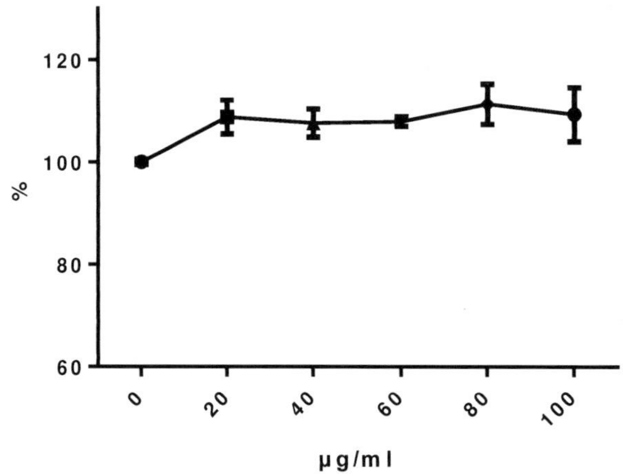

Figure 3. Cell viability after treat with different cerium hollow nanospheres concentrations for 24 hours

The 3-(4, 5-dimethyl-2 thiazoyl) -2, 5-diphenyl-2H-tetrazolium bromide (MTT) assay was used to assess cell viability and evaluate the cytotoxicity in vitro. The MTT results (Figure 3) showed that the cells maintained a high level of viability even at high concentrations. This suggests cerium hollow nanospheres have no inhibition on L02 cells which can provide basis for the further biomedical application.

3.3. Preliminary evaluation as a drug carrier

Erythrocyte protoporphyrin was used as an indicator drug for preliminary testing and evaluation of drug loading and delivery properties of cerium hollow nanospheres by microscopy to observe the fluorescence. Laser confocal scanning microscopy (LCSM) confirmed that the cerium hollow nanospheres and protoporphyrin were uptaken by cells (Figure 4). The red fluorescence emitted by protoporphyrin was clearly visible and full of cytoplasm. It was also clearly be seen that cerium hollow nanospheres loading with protoporphyrin also emitted red fluorescence around the cell and in the cytoplasm. All these have reflected the assessment that there was a potential of cerium hollow nanospheres for fluorescence imaging and drug carrier applications in vitro.

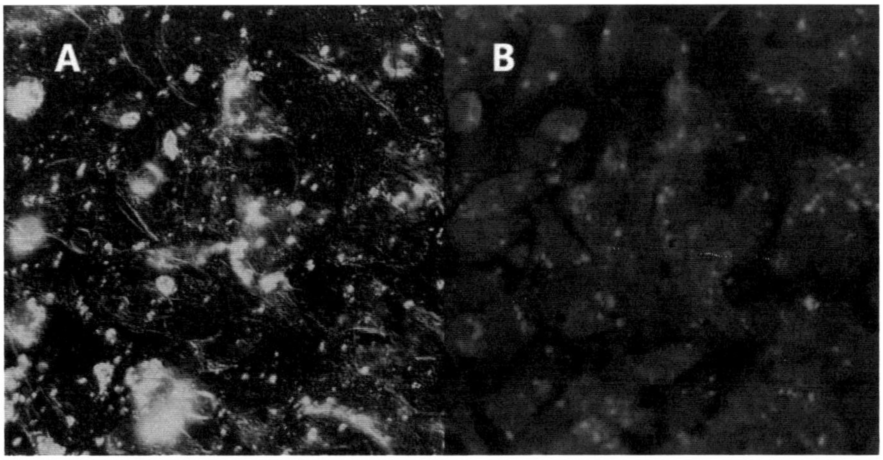

Figure 4. Laser confocal scanning microscopy image: A. Bright field image; B. Fluorescence field image

4. Conclusions
Cerium element has good biological safety and biological activity. The ceria has many excellent properties that it can absorb and block UV, mimic enzymatic activity (superoxide dismutase (SOD) and catalase (CAT)), which may be the main reason makes it possible as a potential drug and might use to fight chronic inflammation and lesions associated with oxidative stress including cancer and neurodegeneration. In this report, cerium hollow nanospheres were used as an exploratory study of drug carriers and as a satisfactory consequence to expand their application in the biomedical field. With the discovery of more biological activity, the scope of application for cerium-based nanomaterials will be greatly expanded in the future, especially in the medical and health care fields.

5. Acknowledgments
This work was financially supported by the National Science Fund for Distinguished Young Scholars of China (No. 81325011).

6. References
[1] Trovarelli A and Fornasiero P, World S F *Catalysis by Ceria and Related Materials* (London: Imperial College Press) p343
[2] Brunelli K, Dabala M, Calliari I and Magrini M 2005 *Corros. Sci.* **47** pp 989-1000
[3] DeCaluwe S C, Grass M E, Zhang C, Gabaly F E, Bluhm H, Liu Z, Jackson G S, McDaniel A H, McCarty K F, Farrow R L, Linne M A, Hussain Z and Eichhorn B W 2010 *J. Phys. Chem.* **114** pp 19853-61
[4] Mori T, Ou D R, Zou J and Drennan J 2012 *Prog. Nat. Sci.-Mater.* **22** pp 561-71
[5] Izu N, Shin W, Matsubara I and Murayama N 2004 *J. Electroceram.* **13** pp 703-6
[6] Meng F, Wang L, Cui J, Li A and Tang H 2015 *Sci. Adv. Mater.* **7** pp 663-9
[7] Tsunekawa S, Sivamohan R, Ohsuna T, Kasuya A, Takahashi H and Tohji K 1999 *Mater. Sci. Forum.* **315** pp 439-45
[8] Deus R C, Foschini C R, Spitova B, Moura F, Longo E and Simoes A Z 2014 *Ceram. Int.* **40** pp 1-9
[9] Nelson B C, Johnson M E, Walker M L, Riley K R and Sims C M 2016 *Antioxidants* **5** pp 152
[10] Caputo F, Nicola, M D and Ghibelli L 2014 *Biochem. Pharmacol.* **92** 112-30
[11] Baker C H 2013 *Transl. Cancer Res.* **2** 343-58

[12] Das S, Dowding J M, Klump K E, McGinnis J F, Self W and Seal S 2013 *Nanomedicine* **8** 1483-1508.
[13] Wason M S and Zhao J 2013 *Am. J. Transl. Res.* **5** 126-31
[14] Li Y, Shi J 2014, *Adv. Mater.* **26**, 3176-205
[15] Zhang C, Li C, Peng C, Chai R., Huang S, Yang D, Cheng Z and Lin J 2010, *Chem. Eur. J.* **16**, pp 5672-80
[16] Qi J, Lai X, Wang J, Tang H, Ren H, Yang Y, Jin Q, Zhang L, Yu R, Ma G, Su Z, Zhao H and Wang D 2015, *Chem. Soc. Rev.* **44**, pp 6749-73
[17] Yuan J, Zhou T and Pu H J 2010, *Phys. Chem. Solids* **71**, pp 1013-9
[18] Guan N, Wang Y, Sun D and Xu J 2009, *Nanotechnology* **20**, 10560310
[19] Das S K, Bhunia M K, Chakraborty D, Khuda-Bukhsh A R and Bhaumik A 2012, *Chem. Commun.* **48**, pp 2891-3
[20] He Q, Shi J 2011, *J. Mater. Chem.* **21**, pp 5845-55
[21] Hughes G A 2005, *Nanomedicine-Nanotechnology* **1**, pp 22-30
[22] Wang S 2009, *Microporous Mesoporous Mater.* **117**, pp 1-9
[23] Tian G, Gu Z, Liu X, Zhou L, Yin W, Yan L, Jin S, Ren W, Xing G, Li S and Zhao Y 2011, *J. Phys. Chem. C* **115**, pp 23790-6
[24] Yang Z, Liu L, Liang H, Yang H and Yang Y 2010, *J. Cryst. Growth* **312**, pp 426-30

Influence of the precursor and annealing temperature on the hydrothermal growth of ZnO nanostructures

R A Rani[1,2], A S Zoolfakar[2], W S W M Sabri[2], S Alrokayan[3], H A Khan[3] and M Rusop[1,2*]**

[1]NANO-SciTech Centre (NST), Institute of Science, Universiti Teknologi MARA, 40450 Shah Alam, Selangor, Malaysia
[2]NANO-ElecTronic Centre (NET), Faculty of Electrical Engineering, Universiti Teknologi MARA, 40450 Shah Alam, Selangor, Malaysia
[3]Research Chair for Biomedical Applications of Nanomaterials, Department of Biochemistry, College of Science, King Saud University, Riyadh 11451, Saudi Arabia

Email: *nanouitm@gmail.com; **ahmad074@salam.uitm.edu.my

Abstract. Wet synthesis of zinc oxide (ZnO) has been reported as an effective and low cost process for large production thin films. The purpose of this project is to investigate the effect of different precursor, different concentration and annealing temperature on the growth of ZnO nanostructures. The ZnO nanostructures is synthesised via a hydrothermal process on a glass substrate and characterized using Field Emission Scanning Electron Microscopy (FESEM). The morphological changes of the nanostructures are systematically investigated as a function of precursors and concentration as well as annealing temperature. The process use two different types of precursor which are Zinc acetate dihydrate and Zinc nitrate hexahydrate.

1. Introduction

ZnO is a unique material that exhibits semiconducting, piezoelectric, pyroelectric properties, excellent chemical and thermal stability, slow electron/hole recombination as well as relatively low cost [1]. Furthermore, ZnO nanostructure with an effective morphology has a significant potential in efficient structure and performance of nanoscale devices due to the quantum restriction of charge carriers in small extent [2].

Several fabrication methods have been reported to produce these morphological structures including electrodeposition, hydrothermal and wet chemical bath deposition (CBD) [3-5]. Among these fabrication methods, the hydrothermal method is the most advantageous because it allows low temperature synthesis, it is environmentally friendly, and it encompasses a large production area . Due to the convenience and simplicity in the operation, hydrothermal method has been proven to be versatile approach for synthesis of ZnO. In this paper, we study the effect of two different types of precursor which are Zinc acetate dihydrate and Zinc nitrate hexahydrate as well as annealing temperature on the hydrothermal growth of nanostructured ZnO.

2. Experimental procedures

Glass substrate was pre-cleaned subsequently in ultrasonic bath of acetone and ethanol. ZnO seed layer were deposited on pre-cleaned glass substrate by spin-coating method. The seed solution is prepared using Zinc Acetate Dihydrate ($Zn(O_2CCH_3)_2(H_2O)_2$), Aluminium Nitrate Hexahydrate

($Zn(NO_3)_2 \cdot 6H_2O$)), Monoethanolamine (MEA) and 2-Metoxyethanol ($C_3H_8O_2$). ZnO seed layer were deposited on pre-cleaned glass substrate by spin-coating method. In the first step, seed solution is prepared using Zinc Acetate Dihydrate ($Zn(O_2CCH_3)_2(H_2O)_2$), Aluminium Nitrate Hexahydrate ($Zn(NO_3)_2 \cdot 6H_2O$)), Monoethanolamine (MEA) and 2-Metoxyethanol ($C_3H_8O_2$). The solution was aged for 24 hours before the spin-coating process. After each spin-coating process, the samples were dried for 10 minutes at 150 °C to evaporate the solvent. The spin-coating and drying process were repeated 5 times to obtain the desired thickness. Finally, the samples were annealed at 500 °C for 1 hour.
The synthesis of ZnO nanostructures *via* hydrothermal process were prepared from aqueous solutions of Zinc Acetate Dihydrate ($Zn(O_2CCH_3)_2(H_2O)_2$) and Hexamethylenetetramine (HMT: $(CH_2)6N_4$). Another aqueous solution used were Zinc Nitrate Hexahydrate ($Zn(NO_3)_2 \cdot 6H_2O$)) and Hexamethylenetetramine (HMT: $(CH_2)6N_4$). These precursors were mixed in a molar ratio of 1:1 in 150 ml distilled water and undergo pre-treatment at 50 °C for 30 minutes. The solution is then vigorously stirred at 250 rpm for 3 hours. After 3 hours of stirring the solution was placed into 150ml flask bottle containing sample. The flask bottle was then immersed in 95 °C water bath for 50 minutes. After 50 minutes, the sample was taken out from the flask bottle and rinsed with distilled water. The sample was then dried at 150 °C for 10 minutes and annealed at 200 °C. The same process was repeated for different annealing temperature which is 500 °C.

3. Result and discussion

3.1. Effect of different concentration

The surface morphology of ZnO nanostructures for different concentration of precursor was studied by FESEM. The samples were studied at different annealing temperature and with different type of precursor. As observed in Figure 1 and Figure 2, the size of ZnO nanostructures increased in increasing concentration and annealing temperature. At a very low concentration, 0.05 M, the size of nanorod diameter was slightly smaller compared to the nanorod grown in 0.1 M precursor solution. Guo *et al.* reported that the low super saturation degree is believed to be responsible for the small-diameter and the narrow size distribution of the ZnO nanorods grown in solutions with low concentration [6]. Wang *et al.* reported that the growth unit diffusing in the solution and adsorbing on the nuclei [7]. The obtained crystal size and shape depend on the nucleation rate as well as the crystal growth rate.
According to Li et. al., different concentration of Zn^{2+} precursor may result in different concentration Zn(OH)2 which is called growth unit [8]. The supply of growth units from $Zn(NO_3)_2 \cdot 6H_2O$ is very little, and then the zinc precursor is transported from the root regions of nanorods to the top regions because of the gradient of concentration. So at the lower concentration regions around the tops of nanorods, the growth rate of ZnO crystals is relatively slow compared with the root of nanorods, which leads to the formation of pyramid like ZnO nanorods. H. Wei *et al.* also conducted an experiment indicates that the diameter of the rods is decreased by decreasing the concentration of the precursor [9].
When the concentration of Zn^{2+} and OH^- exceed saturation, fine ZnO nuclei spontaneously formed in the solution, then the anisotropic nanoparticles of ZnO combine together urged by interfacial free energy as well as the inability of water solution. In other words, it was the "lowest-energy" theory that decided the preferential growing plane. The rule is obeyed strictly throughout the process and no branching is observed in single nanorods. The ultrasonic pretreatment of the solution mixture makes nucleus formation uniform and decreases aggregation [7]. At a higher concentration, 0.4 M, the nanorods grown on the substrate become larger resulting in flakes shape and started to embed each other (Figure 1(c) and (f)). Figure 2((c) and (f)) shows that the nanorods grow in hexagonal shape on the substrate but in hollows and irregular shapes. Guo *et al.* and Zhao *et al.* reported a similar result in which the hollow size increases when the precursor concentration is increased from 0.3 to 0.5 M [10, 11]. The irregular shapes from concentrations of 0.3 and 0.5 M can be explained by the fact that at higher concentrations, the hexagonal rods embed each other, resulting in irregular shapes. Based on

IOP Conf. Series: Materials Science and Engineering **380** (2018) 012018 doi:10.1088/1757-899X/380/1/012018

the image observed from FESEM, it can be concluded that as the concentration of precursor increase, the size of nanorod diameter will also increase.

3.2. *Effect of different precursor*

The exhibited sample using zinc acetate as precursor is observed to be uniformly distributed compared to the sample using zinc nitrate. Morphological structure of sample of using different type of precursor was observed as shown in Figure 1 (zinc nitrate) and Figure 2 (zinc acetate). Figure 1 show that the structure grows uniformly in hexagonal shape in increasing concentration from 0.05 M to 0.1 M but started to embed each other, forming flakes shape at 0.4 M concentration. While in Figure 2, the result shows that the structure grows in nanorods at 0.05 M and start forming dumbbell shape at 0.1 M and 0.4 M but still grow uniformly in hexagonal shape as the concentration increase. The formation of flakes in Figure 1 ((c) and (f)) is due to the acidic strength of zinc nitrate. Singh *et al.* reported that the different morphologies of zinc oxide obtained are because of growth rate difference pertaining to different crystallographic planes and the crystal growth rate depends on the nature of the precursor used [12]. The acidic strength of the zinc oxide precursors used in the present study, which affects the crystal growth rate, is in the following manner: nitrate > acetate. Zinc acetate being lesser acidic than zinc nitrate, resulted in a morphology with flat end hexagonal prisms possessing longer length as compared to the hexagonal prisms produced by zinc nitrate. At higher pH values, hydrolysis/condensation is uncontrolled and unselective which leads to a highly branched structure/polymer chain. At the same time, it will also generate larger interconnected particles/structure [13].

3.3. *Effect of annealing temperature*

For different annealing temperature, the result exhibit that, at higher temperature, 500 °C, the diameter was larger compared to the sample annealed at lower temperature, 200 °C. Figure 1 and 2 shows the comparison of the diameter size of nanorod in which for higher annealing temperature, 500 °C, the size of nanorods are larger than at lower temperature, 200 °C.

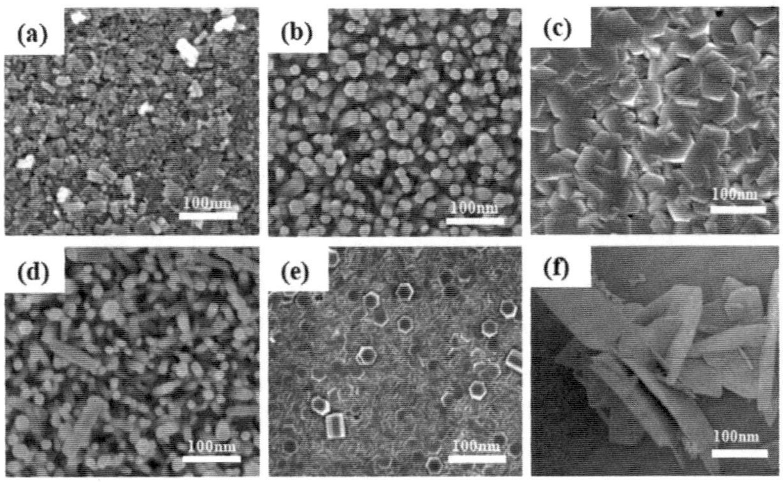

Figure 1. FESEM image of ZnO nanorod prepared using zinc nitrate with different concentration annealed at 200 °C: 0.05 M (a), 0.1 M (b) and 0.4 M (c) and annealed at 500 °C: 0.05 M (d), 0.1 M (e) and 0.4 M (f).

IOP Conf. Series: Materials Science and Engineering **380** (2018) 012018 doi:10.1088/1757-899X/380/1/012018

Figure 2. FESEM image of ZnO nanorod prepared using zinc nitrate with different concentration annealed at 200 °C: 0.05 M (a), 0.1 M (b) and 0.4 M (c) and annealed at 500 °C: 0.05 M (d), 0.1 M (e) and 0.4 M (f).

The condition occurs as at higher temperature, the movement of atom in the substrate increased, thereby changing the structures of ZnO nanorods. Part of nanorods start to aggregate and the average diameter of nanorods start to increase when the annealing temperature increase to 500 °C [14]. Figure 1 shows the rods that grow in smaller size and closer. Unlikely in Figure 2, in which the rod grows larger and apart from each other as the temperature increased. The changes in morphological structure of ZnO using zinc nitrate as precursor can be observed in Figure 1((a), (b) and (c)) at 200 °C and 3((d), (e) and (f)) at 500 °C. The morphological structure of using zinc acetate is shown in Figure 2((a), (b) and (c)) at 200 °C and 2((d), (e) and (f)) at 500 °C. During the annealing process at high temperature, Zhang *et al.* reported that, the microstructure and stoichiometric ratio of materials will change. Due to the effect of annealing treatment, the migration of grain boundaries of ZnO nanorods is stimulated and the crystalline atom gains more activation energy [15]. In this case, the activated atoms will overcome thermal barrier between crystal grains and move freely to the correct sites in the crystal lattice through diffusion. Hence, the rod-like morphology of ZnO will gradually dissolve and convert to another particle with round-edge morphology. If the temperature continuously rises up, the grains with lower surface energy will further grow larger, which results in the final nanoparticle with aggregated morphology [13].

4. Conclusion

This work discusses the morphological structure of ZnO conducted via hydrothermal process. ZnO nanostructured was synthesised using different types of precursor, different concentrations and different annealing temperature. For different types of precursor, zinc acetate result in more uniform size of ZnO nanostructures for each concentration compared to zinc nitrate. In comparison between different concentrations, the best result obtained was by using lowest concentration, 0.05 M, in which it exhibits smaller size of ZnO nanostructure compared to 0.1 M and 0.4 M concentration. Annealing temperature also affect the sizes of ZnO nanorods. The size of ZnO nanorods increased as the annealing temperature increased. At 500 °C, the size of ZnO nanostructures was observed to be bigger than the sample annealed at 200 °C. All samples with single dimension structure and good uniformity of ZnO distribution could be potentially applied in sensor applications.

5. Acknowledgements
This work is supported by Ministry of Education Malaysia (MOE) under the Fundamental Research Grant Scheme (FRGS) (Project code: 600-IRMI/FRGS 5/3 (081/2017)). Special thanks to Research Chair for Biomedical Applications of Nanomaterials, Department of Biochemistry, College of Science, King Saud University, Riyadh 11451, Saudi Arabia.

6. References
[1] S. Wei, S. Wang, Y. Zhang, and M. Zhou, "Different morphologies of ZnO and their ethanol sensing property," *Sensors and Actuators B: Chemical,* vol. 192, pp. 480-487, 2014.
[2] M. A. Kanjwal, F. A. Sheikh, N. A. M. Barakat, X. Li, H. Y. Kim, and I. S. Chronakis, "Zinc oxide's hierarchical nanostructure and its photocatalytic properties," *Applied Surface Science,* vol. 258, pp. 3695-3702, 2012.
[3] A. S. Zoolfakar, R. Ab Kadir, R. A. Rani, S. Balendhran, X. Liu, E. Kats, S. K. Bhargava, M. Bhaskaran, S. Sriram, S. Zhuiykov, A. P. O'Mullane, and K. Kalantar-zadeh, "Engineering electrodeposited ZnO films and their memristive switching performance," *Physical Chemistry Chemical Physics,* vol. 15, pp. 10376-10384, 2013.
[4] N. E. A. Azhar, A. K. Shafura, I. H. H. Affendi, S. S. Shariffudin, I. Saurdi, S. A. H. Alrokayan, H. A. Khan, and M. Rusop, "Investigation of electrical and optical properties of MEH-PPV: ZnO nanocomposite films for OLED applications," *AIP Conference Proceedings,* vol. 1733, p. 020044, 2016.
[5] N. A. Jayah, H. Yahaya, M. R. Mahmood, T. Terasako, K. Yasui, and A. M. Hashim, "High electron mobility and low carrier concentration of hydrothermally grown ZnO thin films on seeded a-plane sapphire at low temperature," *Nanoscale Research Letters,* vol. 10, p. 7, 2015.
[6] M. Guo, P. Diao, and S. Cai, "Hydrothermal growth of well-aligned ZnO nanorod arrays: Dependence of morphology and alignment ordering upon preparing conditions," *Journal of Solid State Chemistry,* vol. 178, pp. 1864-1873, 2005.
[7] H. Wang, J. Xie, K. Yan, and M. Duan, "Growth Mechanism of Different Morphologies of ZnO Crystals Prepared by Hydrothermal Method," *Journal of Materials Science & Technology,* vol. 27, pp. 153-158, 2011.
[8] Z. Li, X. Huang, J. Liu, Y. Li, and G. Li, "Morphology control and transition of ZnO nanorod arrays by a simple hydrothermal method," *Materials Letters,* vol. 62, pp. 1503-1506, 2008.
[9] H. Wei, Y. Wu, N. Lun, and C. Hu, "Hydrothermal synthesis and characterization of ZnO nanorods," *Materials Science and Engineering: A,* vol. 393, pp. 80-82, 2005.
[10] M. Guo, C. Yang, M. Zhang, Y. Zhang, T. Ma, X. Wang, and X. Wang, "Effects of preparing conditions on the electrodeposition of well-aligned ZnO nanorod arrays," *Electrochimica Acta,* vol. 53, pp. 4633-4641, 2008.
[11] X. Zhao, J. Y. Lee, C.-R. Kim, J. Heo, C. M. Shin, J.-Y. Leem, H. Ryu, J.-H. Chang, H. C. Lee, W.-G. Jung, C.-S. Son, B. C. Shin, W.-J. Lee, S. T. Tan, J. Zhao, and X. Sun, "Dependence of the properties of hydrothermally grown ZnO on precursor concentration," *Physica E: Low-dimensional Systems and Nanostructures,* vol. 41, pp. 1423-1426, 2009.
[12] O. Singh, N. Kohli, and R. C. Singh, "Precursor controlled morphology of zinc oxide and its sensing behaviour," *Sensors and Actuators B: Chemical,* vol. 178, pp. 149-154, 2013.
[13] R. Wahab, S. G. Ansari, Y. S. Kim, M. Song, and H.-S. Shin, "The role of pH variation on the growth of zinc oxide nanostructures," *Applied Surface Science,* vol. 255, pp. 4891-4896, 2009.
[14] H. Wu, M. Xue, J. Ou, F. Wang, and W. Li, "Effect of annealing temperature on surface morphology and work function of ZnO nanorod arrays," *Journal of Alloys and Compounds,* vol. 565, pp. 85-89, 2013.
[15] H. Y. Y. Jian Jiao Zhang, Er Jun Guo, Li Ping Wang, Chun Yu Zhang, Jing Chang, Xin Gao, "Effect of Hydrothermal Reaction Conditions on the Sizes and Morphologies of ZnO Nanorods," *Advanced Materials Research,* vol. 765-767, pp. 3170-3175, 2013.

IOP Conf. Series: Materials Science and Engineering **380** (2018) 012019 doi:10.1088/1757-899X/380/1/012019

The effect of zeolite suspension concentration on the formation of seed layer via vacuum seeding prior to NaX zeolite membranes growth

L S M Nazir[1,2], Y F Yeong[1], T L Chew[1], and A Ahmadon[2]

[1]Chemical Engineering Department, Universiti Teknologi PETRONAS, 32610, Bandar Seri Iskandar, Perak, Malaysia
[2]PETRONAS Research Sdn. Bhd, Jalan Ayer Itam, Kawasan Institusi Bangi, 43000 Bandar Baru Bangi, Selangor, Malaysia

Abstract. In the present work, NaX zeolite membranes were synthesized via secondary growth using vacuum seeding method. NaX zeolite suspension solution for vacuum seeding was prepared at various concentration and the resultant seeded supports with homogeneous coverage were selected for hydrothermal synthesis. The resultant membranes were characterized using XRD, SEM and EDX analysis. The XRD results showed that NaX zeolite membranes were successfully formed and the SEM images revealed that continuous zeolite membrane layer was obtained.

1. Introduction

Natural gas often contains impurities such as carbon dioxide (CO_2) and nitrogen (N_2) which need to be removed in order to meet the gas pipeline specification of less than 2 to 4% [1]. CO_2 can cause corrosion in the presence of water and reduces the heating value of natural gas. Conventional processes for the removal of CO_2 usually use amine absorption and polymeric membrane separation. However, processing high pressure natural gas can be expensive for amine absorption [2] and polymeric membranes are known to be susceptible to plasticization by CO_2 resulting in reduction of selectivity [3]. Porous inorganic membrane such as zeolite membrane which contain nano pores, are favourable for CO_2 separation from CH_4 because of their chemical resistance to CO_2 induced plasticization and superior selectivity compared to polymeric membranes [2,4-5]. N_2 presence in natural gas also reduces the heating value of the gas, and thus required removal. Cryogenic technology is another mature technology used for N_2 removal from natural gas. However, this technology is expensive and energy intensive. Although polymeric membrane have been used for N_2 removal from CH_4 [6], its performance is low compared to the zeolite membrane.

Among all types of zeolites, the most interest for zeolite membrane studies has been focused on these 8 types of zeolite framework i.e., MFI, LTA, FAU, CHA, T, DDR, MEL, and AEI [7]. FAU-type zeolite which includes NaX (Si/Al ratio of 1-1.5) and NaY (Si/Al ratio >1.5) is within the large pore zeolite group with 12-membered rings and of approximately 0.74 nm in diameter. Because of its large pores, FAU membranes offer high CO_2 flux at reasonable selectivity. It is able to separate CO_2 and N_2 from other gases due to its strong absorption affinity resulting in high selectivity for CO_2/CH_4 and moderate selectivity for N_2/CH_4 gases [5,8]. Another important advantage of FAU zeolites is that the preparation does not always require organic structure directing agent (SDAs), thus the cost of fabrication is cheaper compared to the other types of zeolite membrane. In addition, the removal of

SDA via calcination at high temperature is not required and thus, less energy are involved during the fabrication process [7]. Besides, heat induced cracks on the prepared membrane can be eliminated. Successful CO_2/CH_4, CO_2/N_2, CO_2/H_2 and N_2/CH_4 separation by FAU membrane have been reported [9-12].

One of the synthesis method that is widely applied for the preparation of zeolite membrane is secondary growth method where a layer of zeolite seeds is deposited on the membrane support prior to hydrothermal growth [13]. Similar techniques are also applied for the synthesis of FAU zeolite membrane. There are two challenging issues related to the synthesis of FAU zeolite membrane by secondary growth method, which are the preparation of colloidal suspension of zeolite particles, and homogeneous deposition of seeds onto the support surface [14]. Dip coating method is the most commonly used technique for seeding. However, it is reported that continuous seed layers are difficult to achieve as the zeolite seeds cannot closely attach to the support and the colloidal suspensions easily dribble when the supports are withdrawn [14]. Vacuum coating is another seeding method which has been applied by various researchers for the preparation of seed layer. This technique involved dead-end filtration and cross flow filtration method [15]. Vacuum seeding via dead end or cross flow mode can potentially allow a complete coverage and homogeneous distribution of seeds on the support surface [15]. However, proper technique and parameters are required in order to avoid excessive crystals accumulation and thus, uniform coverage of seed on the support surface can be obtained.

In this paper, synthesis of NaX zeolite membrane via secondary growth using vacuum seeding is reported. The effect of different zeolite suspension concentrations on the formation of seed layer via vacuum seeding method prior to the growth of NaX zeolite membranes is studied.

2. Experimental

2.1. Preparation of NaX zeolite seed

NaX zeolite powder was synthesized by insitu crystallization following procedure as described in literature [5]. The solution for the synthesis of zeolite powder was prepared by mixing sodium silicate, sodium aluminate, sodium hydroxide (NaOH) and deionized (DI) water with composition of Al_2O_3 : $4.8SiO_2$: $17Na_2O$: $975H_2O$. Then, the solution was stirred for 4 hours in order to age the mixture. The synthesis gel was filled into autoclave and hydrothermal synthesis was carried out at 90°C for 24 hours. After synthesis, the resultant crystals were filtered and washed with DI water as many times as required to reduce the filtrate pH to less than 10. The filtered crystal was collected and dried in oven at 90 °C.

2.2. Preparation of zeolite seed layer and synthesis of NaX zeolite membrane

For the preparation of zeolite seed layer, a suspension of zeolite crystal was prepared by dispersing between 0.25 to 1.0 g NaX zeolite powder in 100 to 200 mL deionized water followed by an ultrasonic treatment for 60 minutes in order to avoid the aggregation of particles. Four different zeolite suspension solutions with concentration of 0.25 g/100 mL, 0.5 g/100 mL, 0.5 g/200 mL and 1.0 g/200 mL were prepared. The seed was attached onto the disc-type support via dead end vacuum filtration method. Table 1 shows the seeded supports prepared in the present work.

Table 1. Seeded supports prepared using different concentrations of zeolite suspension.

Seeded Support	Concentration of Zeolite Suspension	Seeding Duration (minutes)
NaX-sA	0.25 g/ 100 mL	3
NaX-sB	0.50 g/ 100 mL	3
NaX-sC	0.50 g/ 200 mL	6
NaX-sD	1.00 g/ 200 mL	6

IOP Conf. Series: Materials Science and Engineering **380** (2018) 012019 doi:10.1088/1757-899X/380/1/012019

The NaX zeolite membrane was synthesized using the same synthesis solution and hydrothermal synthesis parameters as those used for the preparation of NaX particles. Seeded disc support was placed in autoclave filled with synthesis solution prepared using the same molar composition that were used for the synthesis of NaX zeolite powder. Hydrothermal synthesis was carried out at 90°C for 24 hours. After synthesis, the membrane was washed with DI water and dried in air at ambient temperature.
The morphology of NaX zeolite membranes was determined by using scanning electron microscopy, SEM (Hitachi TM 3000). The powder samples collected were characterized using X-ray diffraction (XRD) and energy dispersive X-ray (EDX) analysis was used to confirm the Si/Al ratio of the sample.

3. Results and discussion

3.1. Effect of vacuum seeding solution concentration on the formation of seeded layer
Figure 1 shows the SEM images of the distribution of zeolite seed layer on disc support surface using different suspension solution concentration. Referring to Figure 1(a) and 1(c), it can be observed that low to medium zeolite seed coverage are obtained. From the SEM images shown in Figure 1, zeolite suspension solution with concentration of 0.5 g/100 mL and 1.0 g/200 mL resulted in homogenous coverage of seeded layer (Figure 1 (b) and 1 (d)). Both of these support layer (NaX-sB and NaX-sD) are seeded with the same zeolite suspension solution concentration, but with different volume and amount of seed used, as well as the seeding duration.

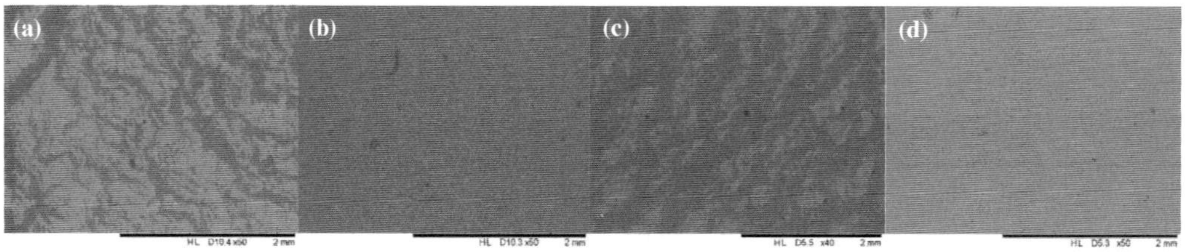

Figure 1. Distribution of zeolite seed layer on disc support surface using different zeolite suspension concentration of (a) 0.25 g/100ml (NaX-sA), (b) 0.5 g/100ml (NaX-sB), (c) 0.5 g/200ml (NaX-sC), and (d) 1.0 g/200ml (NaX-sD).

3.2. Formation of NaX zeolite membrane
Seeded supports with homogeneous seed layer, NaX-sB and NaX-sD were proceeded to NaX zeolite membrane growth via hydrothermal method. Figure 2 shows the SEM images of NaX zeolite membrane (NaX-B) growth on NaX-sB while Figure 3 shows the SEM images of NaX membrane (NaX-D) growth on NaX-sD. From Figure 2, it can be observed that the NaX zeolite membrane grown on NaX-sB exhibited full coverage and continuous zeolite membrane layer.
On the other hand, referring to Figure 3, non-homogeneous growth of NaX-D zeolite membrane is observed. This is mainly because of the zeolite seed layer coated during vacuum seeding was detached from the support surface. NaX-sD was coated with double the duration of NaX seed compared to NaX-sB and thus, resulted in thicker zeolite seed layer. According to Huang et al. [14] coating time and the zeolite seed suspensions concentration could affect the properties of the seeded layer and thus, the formation of the zeolite membrane. Longer coating duration causes the seeded layer to become too thick and crack can easily form which resulting in poor zeolite membrane formation. On the contrary, shorter coating time produced non-continuous zeolite seed layer which could also affect the homogeneous growth of the zeolite membrane. In this case, a thicker zeolite seed layer was easily detached from membrane support during hydrothermal synthesis which resulted in membrane defects. Overall, from the result obtained in this work, zeolite suspension solution with concentration of 0.5 g/100 mL and coating duration of 3 minutes were found to be the most suitable seed solution concentration for vacuum seeding in order to obtain well intergrowth zeolite layer.

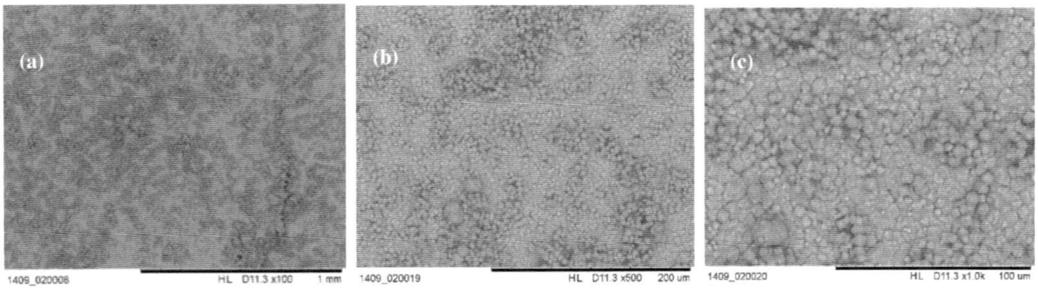

Figure 2. NaX-B zeolite membrane grown on seeded support formed using 0.5 g/100 mL of zeolite suspension solution at (a) 100x magnification, (b) 500x magnification and (c) 1000x magnification.

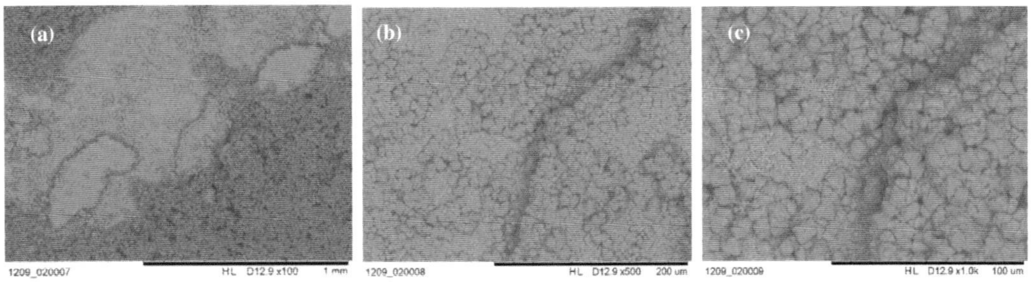

Figure 3. NaX-D zeolite membrane grown on seeded support formed using 1.0 g/200 mL of zeolite suspension solution at (a) 100x magnification, (b) 500x magnification and (c) 1000x magnification.

Figure 4 shows the XRD results for zeolite powder collected after hydrothermal synthesis of NaX-B and NaX-D zeolite membranes. The XRD pattern obtained was also compared with the XRD pattern obtained for the commercial zeolite NaX (Tosoh Corp., Si/Al ratio of 1.25). It can be seen from Figure 4 that the same zeolite phases were produced among these three samples. The resulting Si/Al ratio for NaX-B and NaX-D are are 1.28 and 1.36, respectively, which confirmed the formation of NaX zeolite.

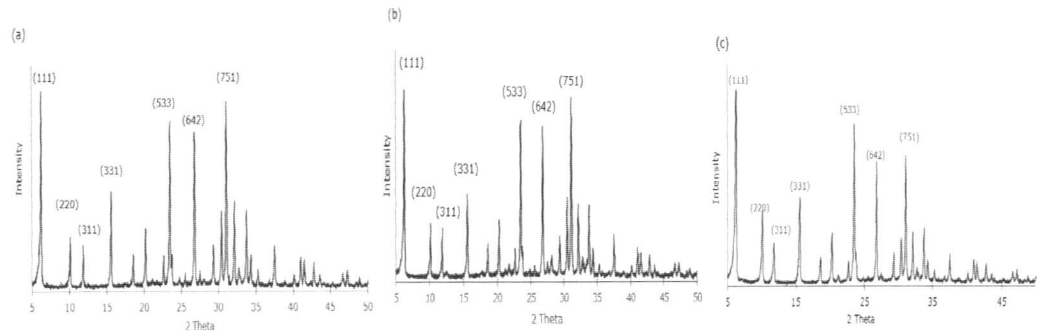

Figure 4. XRD results for NaX zeolite (a) NaX-B, (b) NaX-D and (c) Commercial NaX.

4. Conclusion

In this work, NaX zeolite membrane have been synthesized via secondary growth using vacuum seeding method and the structure of the zeolite membranes have been verified by XRD analysis. The effect of zeolite suspension concentration on the formation of seed layer prior to membrane growth was studied. Zeolite suspension solution with concentration of 0.5 g/100 mL was found to be the most suitable seed solution concentration for vacuum seeding in order to obtain well intergrowth zeolite layer. However, it is noted that, this concentration is suitable for the surface area of membrane support used for this study. Membranes with different (higher or lower) surface area might need different concentration of zeolite suspension solution that is suitable to achieve thin and homogenous coverage zeolite seed layer. Further work will be focusing on the gas permeation testing of the membrane in CO_2/CH_4, CO_2/N_2 as well as N_2/CH_4 separation.

5. Acknowledgment

The financial support provided by PETRONAS and the technical supports provided by Institute of Contaminant Removal and PETRONAS are highly appreciated.

6. References

[1] A. J. Kidnay and W. R. Parrish, "Fundamentals of natural gas processing mechanical engineering." *Mech. Eng.*, p. 418, 2006.
[2] T. Wu, M. C. Diaz, Y. Zheng, R. Zhou, H. H. Funke, J. L. Falconer, and R. D. Noble, "Influence of propane on CO_2/CH_4 and N_2/CH_4 separations in CHA zeolite membranes," *J. Memb. Sci.*, vol. 473, pp. 201–209, 2015.
[3] R. Swaidan, B. Ghanem, E. Litwiller, and I. Pinnau, "Physical Aging, Plasticization and Their Eff ects on Gas Permeation in 'Rigid' Polymers of Intrinsic Microporosity," *Macromolecules*, no. i, 2015.
[4] T. Tomita, K. Nakayama, and H. Sakai, "Gas separation characteristics of DDR type zeolite membrane," *Microporous Mesoporous Mater.*, vol. 68, pp. 71–75, 2004.
[5] S. Li, J. L. Falconer, and R. D. Noble, "SAPO-34 membranes for CO_2/CH_4 separations: Effect of Si/Al ratio," *Microporous Mesoporous Mater.*, vol. 110, no. 2–3, pp. 310–317, 2008.
[6] K. A. Lokhandwala, "Nitrogen Removal from Natural Gas," 1999.N. Kosinov, J. Gascon, F. Kapteijn, and E. J. M. Hensen, "Recent developments in zeolite membranes for gas separation," *J. Memb. Sci.*, vol. 499, pp. 65–79, 2016.
[7] N. Kosinov, J. Gascon, F. Kapteijn, and E. J. M. Hensen, "Recent developments in zeolite membranes for gas separation," *J. Memb. Sci.*, vol. 499, pp. 65–79, 2016.
[8] Y. Han, J. Kang, H. Kim, J. Moon, C. Cho, and C. Lee, "Separation of Carbon Dioxide and Methane Mixture by an Adsorbent / Membrane Hybrid System Using Zeolite 5A Pellets and FAU-Zeolite Membrane," 2017.
[9] Y. Hasegawa, T. Tanaka, K. Watanabe, and B. Jeong, "Separation of CO_2 - CH_4 and CO_2 - N_2 Systems Using Ion-exchanged FAU-type Zeolite Membranes with Different Si/Al Ratios," Korean *J. Chem. Eng.*, vol. 19, no. 2, pp. 309–313, 2002.
[10] X. Gu, J. Dong, and T. M. Nenoff, "Synthesis of defect-free FAU-type zeolite membranes and separation for dry and moist CO_2/N_2 mixtures," *Ind. Eng. Chem. Res.*, vol. 44, no. 4, pp. 937–944, 2005.
[11] K. Kusakabe, T. Kuroda, and S. Morooka, "Separation of carbon dioxide from nitrogen using ion-exchanged faujasite-type zeolite membranes formed on porous support tubes," *J. Memb. Sci.*, vol. 148, pp. 13–23, 1998.
[12] K. Weh, M. Noack, I. Sieber, and J. Caro, "Permeation of single gases and gas mixtures through faujasite-type molecular sieve membranes," *Microporous Mesoporous Mater.*, vol. 54, pp. 27–36, 2002.
[13] P. Gorgojo, Ó. de la Iglesia, and J. Coronas, "Preparation and Characterization of Zeolite Membranes," *Membr. Sci. Technol.*, vol. 13, no. 7, pp. 135–175, 2008.
[14] A. Huang, Y. S. Lin, and W. Yang, "Synthesis and properties of A-type zeolite membranes by secondary growth method with vacuum seeding," *J. Memb. Sci.*, vol. 245, pp. 41–51, 2004.

IOP Conf. Series: Materials Science and Engineering **380** (2018) 012019 doi:10.1088/1757-899X/380/1/012019

[15] C. Algieri, P. Bernardo, G. Barbieri, and E. Drioli, "A novel seeding procedure for preparing tubular NaY zeolite membranes," *Microporous Mesoporous Mater.*, vol. 119, pp. 129–136, 2009.

IOP Conf. Series: Materials Science and Engineering **380** (2018) 012020 doi:10.1088/1757-899X/380/1/012020

Hydraulic flow units for reservoir characterization: A successful application on arab-d carbonate

H Abdulelah[1], S Mahmood[1] and G Hamada[1]

[1]Department of Petroleum Engineering, Universiti Teknologi PETRONAS, 32610 Seri Iskandar, Perak Darul Ridzuan Malaysia

Abstract. The characterization of carbonate formations is challenging as compared to sandstones, yet carbonate reservoirs hold over 60% of the world's hydrocarbon reserves. Carbonate reservoirs exhibit a high level of heterogeneity at every scale; from core to field. To be able to manage heterogeneity for reservoir modelling, the formation has to be discretized into a few rock types, each of which having somewhat similar flow properties. Recently, the interest in extending the rock-typing approaches is increasing with the aim to identify the potential layers in complex lithology like carbonates. The approach becomes more rigorous if the geological description is coordinated with petrophysical data, an approach that has been followed in this study. The hydraulic flow units in Arab-D formation were identified and interpreted using both geological facies and petrophysical data. All three methods; histogram analysis, normal probability plot and least-squared regression were utilized to determine the optimum number of hydraulic flow units across Arab-D carbonate formation. Published routine core analysis data from ten wells of Arab-D formation was analyzed and six optimum hydraulic flow units were identified. The average porosity and average permeability of each hydraulic flow unit was then computed. The results were found to be in good agreement with the geological facies data of the Arab-D formation, thus validating the identified flow units.

1. Introduction

Characterization of carbonate reservoirs is challenging due to The existence of heterogeneities at each scale caused by diagenesis [1]. In the past years, there have been many attempts to establish a universal workflow to characterize carbonates using both geological and petrophysical data. However, many of these approaches are not rigorous enough in integrating all data and are inadequate for up-scaling [2]. Hydraulic flow units (HFU) is one of the most widely used to accurately describe and characterize carbonate reservoirs [3-5]

Hydraulic flow unit (HFU) is defined as a rock volume unit whose members have somewhat similar flow properties which are distinct from properties in the other units. The concept of HFU is different than lithofacies as it aims to cluster similar fluid pathway in the reservoir rather than on the basis of distribution of lithologies [6]. HFU can be determined by making use of rock quality index, void ratio and flow zone indicator (Equations 1 to 3) which also relate to the mean hydraulic radius [7, 8]. The reservoir quality index (RQI) is defined as the ratio of permeability to porosity:

$$RQI = 0.0314\sqrt{\frac{k}{\varphi}} \quad (1)$$

The void ratio or the normalized porosity, φ_z is defined as:

$$\varphi_z = \frac{\varphi}{1-\varphi} \tag{2}$$

By excluding the microscopic measurement parameters and lumping the Carmen-Kozeny parameters, a variable called Flow zone indicator (FZI) can be introduced, which is a function of reservoir quality index (RQI) and void ratio:

$$FZI = \frac{RQI}{\varphi_z} \tag{3}$$

Where RQI is the rock quality index (μm), k is permeability (mD), is effective porosity (v/v), φ_Z is normalized porosity, and FZI is the flow zone index (μm).
By taking log on both sides of Equation. 3 and re-arranging:

$$\text{Log RQI} = \log \text{FZI} + \log\ \varphi_z \tag{4}$$

Equation. 4 suggests that log-log plot of RQI versus φ_Z should yield straight line of which the slope is one and the intercept equals FZI. Rocks that falls on the same straight line are considered to be in the same hydraulic flow unit, with a unique value of FZI obtained from the intercept.
This concept has been employed by many researchers to characterize giant carbonate reservoirs in USA, Russia, and Middle East [9-12]. However, none of these studies had linked the petrophysical properties hydraulic flow units (HFUs) to their corresponding geological data. In this study, the concept of hydraulic flow unit (HFU) is implemented to characterize the prolific Arab-D carbonate formation. In doing so, around 480 published routine core analysis data (RCA) were collected and used for this purpose [13]. The RCA data included porosity, permeability, grain matrix density, and facies descriptions for 480 core plugs.

2. Geological background

Figure 1. shows Ghawar field which is the world's largest oilfield located in Saudi Arabia at about 200 km east of Riyadh.

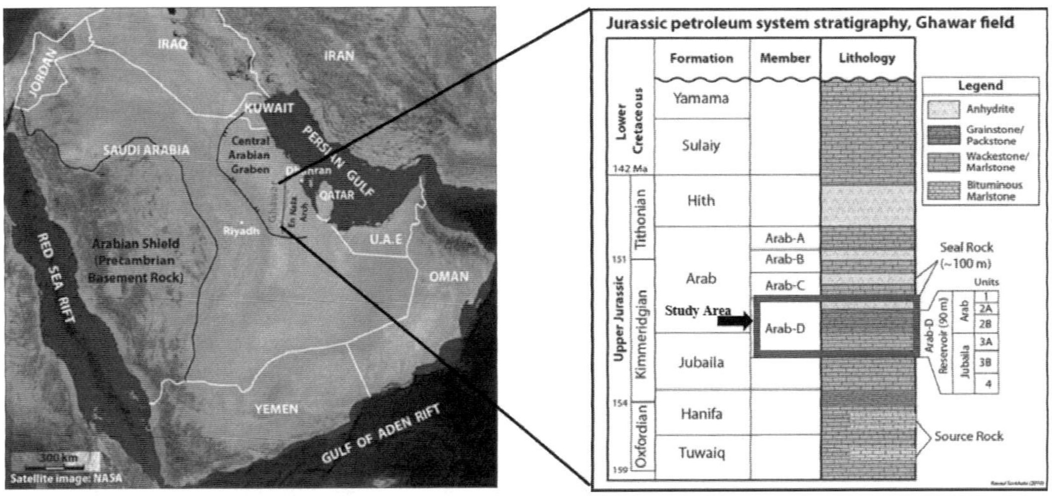

Figure 1. Location of Ghawar field and study area [11].

The field is 280 km long and 30 km wide on average. It is divided into six areas which are Fazran, Ain Dar, Shedgum, uthmaniyah, Haradh and Hawiyah from North to South. The field covered with mainly sediments from Miocene-Pleistocene with two small outcrops of Eocene. Upper Jurassic Arab formation constitutes of four main members (from bottom to top): Arab-D, Arab-C, Arab-B, and Arab-A. The main reservoir in Ghawar field is the carbonate part of the Arab-D. The Arab D formation is mainly made up of packestone and grainstone with underlying formation of Jubaila which consists of limestone and mudstones with an average porosity of 15%. Dolomites also occur in the formation with very high permeability contributing to high production rates. Along Arab-D formation, six lithofacies were reported; namely Skeletal Oolitic Above Sequence Boundary, Cladocoropsis, Stromatoporoid-Red Algae-Coral, Skeletal Oolitic Below Sequence Boundary, Bivalve-Coated Grain-Intraclast and Micrite [13, 14].

3. Methodology

The optimum number of hydraulic flow units (HFU) across Arab-D formation were identified using the 480 routine core analysis (RCA) data. As in Figure 2, Histogram analysis, probability analysis and linear regression were the three methods employed for this purpose. Histogram analysis of flow zone indicator (FZI) data was first implemented to get a quick identification of the number of existing hydraulic flow units. The FZI data were then analyzed using probability analysis method. Lastly, several iterations of the linear regression analysis between rock quality index (RQI) and normalized porosity (φ_Z) were conducted to find the optimum number of HFUs. The correlation coefficients of regression (R^2) were calculated and the optimum number of HFU's were deduced when no significant change in correlation coefficients between RQI and φ_Z occurred (Figure 5). The porosity and permeability data belonging to each HFU was then averaged to represent its k-φ value. To validate the findings, the characteristics of the arbitrarily identified HFU's were correlated against the geological facies data for the reservoir and a good agreement was found.

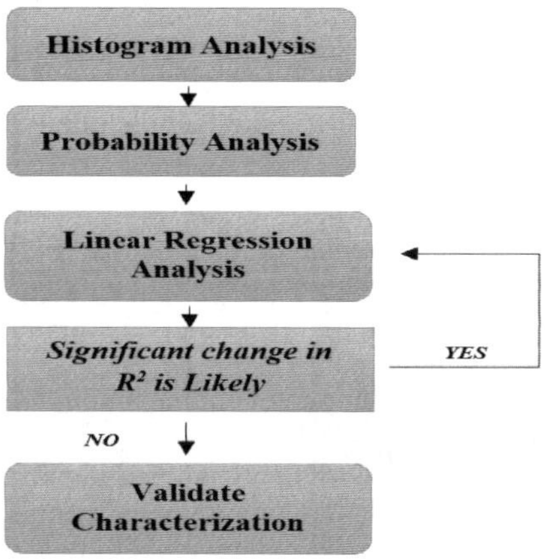

Figure 2. Work flow chart for finding optimum rock clusters based on HFUs.

The following sections will further elaborate the three method.

3.1. Histogram analysis method

The histogram analysis method serves as a primary method to identify the number of HFU's existing in a heterogeneous carbonates reservoirs. Since FZI is an indirect function of permeability, a histogram of log FZI would reveals "n" numbers of normal distributions corresponding to "n" numbers of HFU's. In this research, the flow zone index (FZI) values were first calculated using equation 2. A histogram plot for FZI value is then constructed. The number of HFU's corresponds to the number of normal distributions in the histogram plot of Log (FZI). A new FZI is determined to represent each HFU by averaging all the FZI within the same HFU.

IOP Conf. Series: Materials Science and Engineering **380** (2018) 012020 doi:10.1088/1757-899X/380/1/012020

3.2. Probability analysis method

The probability method is applied for more investigation. The data of Log FZI was plotted versus the percentile in a normal probability plot. The number of optimum HFU's was revealed by the number of significant straight lines observed in the normal probability plot.

3.3. Least square regression method

Least square regression method was also utilized to identify the optimum number of existing HFU's. Rock quality index (RQI) and normalized porosity (φ_Z) values were calculated using equation 1 and equation 3. In a log-log scale, a plot of RQI versus φ_Z was then constructed. Initially, it was assumed that only one HFU exists and the coefficient of determination (R^2) was calculated. In the same way, the process was repeated by assuming a higher number of HFU's until there is no significant change in R^2 values. In this method, each HFU is represented by straight line with unit slope in the log-log plot of RQI versus φ_Z. Once the optimum number of HFU's were confirmed, the k-ϕ transforms were established and validated with the geological information of cores.

4. Results & discussion

4.1. Optimum number of hydraulic flow units

Three methods were used to identify the optimum number of hydraulic flow units (HFU) that can be presumed to exist in the "Arab D" formation. The analyses using each of these methods is presented here along with a comparison of their efficacy.

4.1.1. Analysis by histogram method

Figure 3 is histogram plot of log FZI as plotted using method explained in section 3.1. It is obvious from the plot that more than one flow units exists in Arab-D formation. However, it is difficult to identify the exact number of flow units existing in this formation because the histogram plot of log FZI is a superposition of a number of overlapping individual normal distributions. Unless the number of individual distributions is isolated/identified, one cannot predict the number of existing hydraulic flow units. Since the accuracy of this method is influenced by transition zones between HFU's, the analysis of "Arab D" flow units using histogram method is a qualitative one that show only that a variation in the distribution of HFU's across Arab-D formation exists.

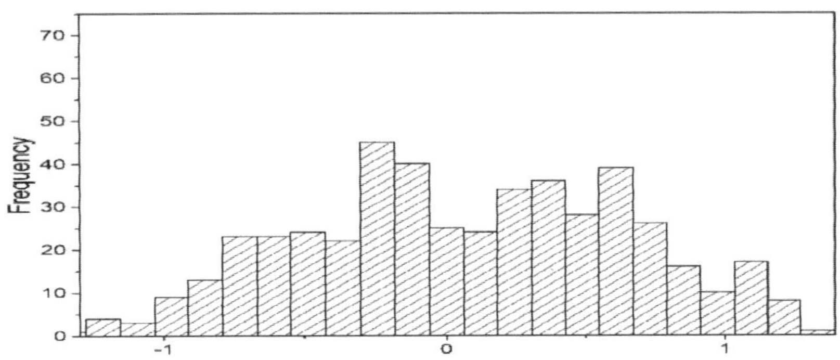

Figure 3. Histogram plot of log FZI (Flow Zone Indicator).

4.1.2. Analysis by probability plot method

IOP Conf. Series: Materials Science and Engineering **380** (2018) 012020 doi:10.1088/1757-899X/380/1/012020

The results of this method are quantitative and more accurate than histogram method. Figure 4 is the normal probability plot of log FZI. Six distinct straight lines can be identified when log FZI is plotted on a log normal probability plot which suggests the existence of six HFU's along Arab-D formation.

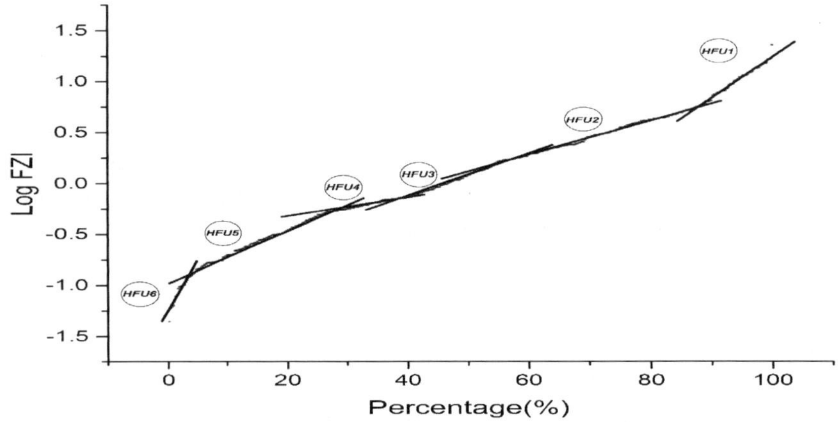

Figure 4. Normal probability plot of log FZI

4.1.3. Analysis by least-square regression method

This is a more rigorous/accurate quantitative method for establishing optimum number of HFUs and determining the porosity-permeability relationships for each identified HFU. Figure 5 was constructed to unravel the optimum number of existing HFUs. Initially, one single HFU was assumed to exist in the whole Arab –D formation. The permeability–porosity relationship was then established and the predicted permeability was correlated with the permeability data from routine core analysis (RCA) and the value of coefficient of determination (denoted by R^2) was determined by regression analysis. R^2 value ranges between 0 and 1 indicating the extent to which the dependent variable is predictable. The R^2 determined for 1-HFU assumption for Arab-D was 0.75, which indicates a fair degree of correlation. The above process was repeated by increasing the number of HFUs by one and determining the R^2 for the corresponding number of HFU. The two R^2 values were plotted in a R^2 vs Number of HFUs graph. This process was continued until increasing the number of HFU further did not improve the R^2 value i.e. the curve leveled off. A careful examination of the completed plot (Figure 5) reveals that the curve levels off when the number of HFUs is six. Thus, considering number of HFUs more than six will not improve the analysis and six is the optimum number.

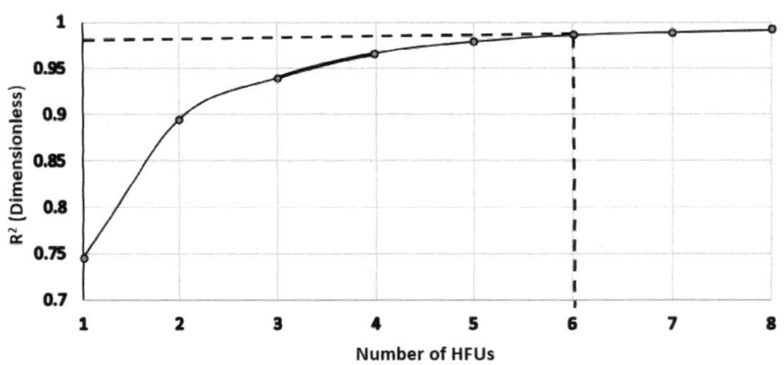

Figure 5. Coefficient of determination versus number of HFU's

IOP Conf. Series: Materials Science and Engineering **380** (2018) 012020 doi:10.1088/1757-899X/380/1/012020

To determine the porosity-permeability relationship of each HFU, the reservoir quality index (RQI) was calculated using the porosity-permeability data from core analysis and applying Equation. 1. Also, the porosity data was normalized by using Equation. 3 and is denoted by φ_Z. The FZI value for each data point was calculated using Equation. 3. The data was clustered in to six (the optimum number of HFUs) by choosing appropriate ranges of FZIs for each HFU. The average FZI was then computed for each HFU cluster. Figure 6 shows the RQI versus normalized porosity (φ_Z) data plotted on a log-log scale. To draw a line for each HFU, a straight line of slope 1 was constructed and pinned at the corresponding average FZI as intercept (the intersection of straight line at $\varphi_Z = 1$ since it is a log-log plot).

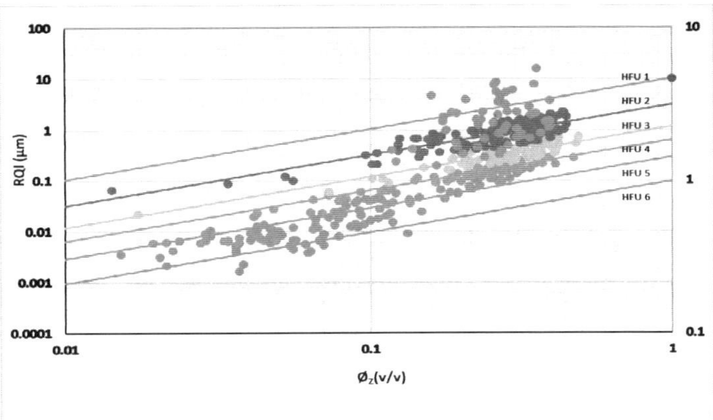

Figure 6. Reservoir quality index vs normalized porosity

The RQI value of the HFUs is read at the intersection of the straight line to RQI scale at the origin. Having the HFUs identified and verified by Figure 7, the average porosity and average permeability values were determined by taking an average of all the data belonging to an HFU. Table. 1 shows the predicted porosity and permeability values for each of the determined flow units.

Table 1. Average porosity and permeability of predicted HFU's.

HFU#	Average Permeablity (mD)	Average Porosity (%)
1	1789	19
2	451	22
3	90	22
4	10	19
5	0.333	9
6	0.009	6

4.1.4. Validating the characterization

To validate the results, the predicted HFU's were combined with the existing lithofacies' description from the Rosetta stone project. For example, Figure 6 indicates that the dominant Lithofacies in HFU 1 are Skeletal-oolitic packstone, which was reported to possesses a horizontal permeability up to 1451 (mD) in a core scale[13]. The poor permeability in HFU 5 and HFU 6 is attributed to micrite which was found to be the abundant lithofacies existing in those two hydraulic flow units. Thus, these results are in good general agreement with the predicted properties of hydraulic flow units as in Table 1.

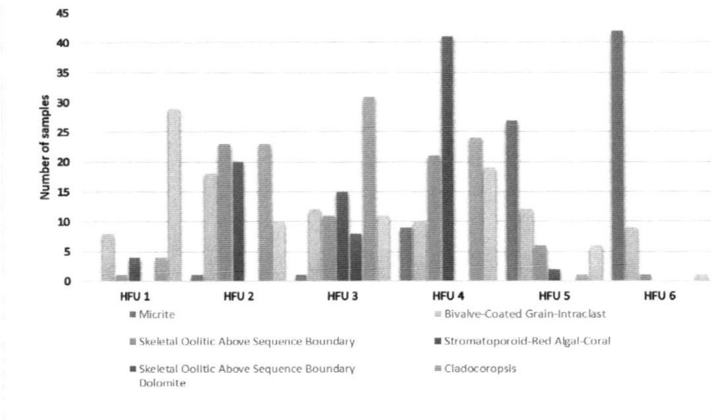

Figure 7. Litholofacies of each HFU.

5. Conclusions

Based on the case study of Arab D" formation, following observations/conclusion can be made:

- Clustering the rocks into families with similar flow properties (HFU) is a viable tool for reservoir characterization of carbonate reservoirs.
- Histogram analysis of FZI data is a qualitative approach but has limitations since it can't identify the optimum number of HFU's objectively, thus its use as a stand-alone technique be avoided.
- Probability analysis of FZI data is a quantitative approach which gives satisfactory results. However, it should be used in conjunction with the "Least Square Regression" to improve the confidence level.
- "Least Square Regression" method is probably the most rigorous method for determining the HFUs which can be used as a standalone method.
- For future studies, artificial intelligence and data science tools should be implemented for such characterizations. This would be more efficient.

6. References

[1] A. S. Bagci and C. Y. Akbas, "Permeability Estimation Using Hydraulic Flow Units in Carbonate Reservoirs," in *Rocky Mountain Oil & Gas Technology Symposium*, 2007.

[2] M. Rebelle, "Rock-typing In Carbonates: A Critical Review Of Clustering Methods," in *Abu Dhabi International Petroleum Exhibition and Conference*, 2014.

[3] J. Radiansyah, T. E. Putra, R. Ismail, R. A. Wibowo, E. E. Riza, and M. Kurniawan, "Reservoir Description using Hydraulic Flow Unit and Petrophysical Rock Type of PMT Carbonate Early Miocene of Baturaja Formation, South Sumatra Basin," in Adapted from extended abstract prepared in conjunction with presentation *AAPG International Conference & Exhibition*, Istanbul, Turkey, 2014.

[4] R. Mohebian, M. A. Riahi, and A. Kadkhodaie, "Characterization of hydraulic flow units from seismic attributes and well data based on a new fuzzy procedure using ANFIS and FCM algorithms, example from an Iranian carbonate reservoir," *Carbonates and Evaporites,* pp. 1-10, 2017.
[5] S. K. Mahjour, M. K. G. Al-Askari, and M. Masihi, "Flow-units verification, using statistical zonation and application of Stratigraphic Modified Lorenz Plot in Tabnak gas field," *Egyptian Journal of Petroleum,* vol. 25, pp. 215-220, 2016.
[6] W. Ebanks Jr, "Flow unit concept-integrated approach to reservoir description for engineering projects," *AAPG (Am. Assoc. Pet. Geol.) Bull.;(United States),* vol. 71, 1987.
[7] A. A. Abed, "Hydraulic flow units and permeability prediction in a carbonate reservoir, Southern Iraq from well log data using non-parametric correlation," *Int J Enhanc Res Sci Technol Eng,* vol. 3, pp. 480-486, 2014.
[8] N. Bize-Forest, V. Baines, A. Boyd, A. Moss, and R. Oliveira, "Carbonate Reservoir Rock Typing and the Link between Routine Core Analysis and Special Core Analysis," in *International Symposium of the Society of Core Analysts*, 2014, pp. 8-11.
[9] D. A. Alobaidi, "Permeability Prediction in One of Iraqi Carbonate Reservoir Using Hydraulic Flow Units and Neural Networks," 2016.
[10] O. Deghirmandjian, "Identification and characterization of Hydraulic Flow Units in the San Juan Formation, Orocual Field, Venezuela," Texas A&M University, 2001.
[11] F. A. Al-Ajmi and S. A. Holditch, "Permeability Estimation Using Hydraulic Flow Units in a Central Arabia Reservoir," 2000.
[12] A. Kassenov and Z. Uzykanov, "Application of Hydraulic Flow Units for History Matching of the Hydrodynamic Model of Tengiz Field," in *SPE Annual Caspian Technical Conference & Exhibition*, 2016.
[13] E. A. Clerke, H. Mueller, E. C. Phillips, R. Y. Eyvazzadeh, D. H. Jones, R. Ramamoorthy, *et al.*, "Application of Thomeer Hyperbolas to decode the pore systems, facies and reservoir properties of the Upper Jurassic Arab D Limestone, Ghawar field, Saudi Arabia: A "Rosetta Stone" approach," *GeoArabia,* vol. 13, pp. 113-160, 2008.
[14] R. Sorkhabi. (2010, September 2010) The King of Giant Fields. *GEO ExPro* [Geoscience]. 88.

IOP Conf. Series: Materials Science and Engineering **380** (2018) 012021 doi:10.1088/1757-899X/380/1/012021

Application of nanomaterials to enhanced the lubricity and rheological properties of water based drilling fluid

A R Ismail[1,2], M S A Rashid[2] and B Thameem[2]

[1]MPRC-UTM Institute for Oil & Gas, 81310 Johor Bahru, Malaysia
[2]Petroleum Eng. Dept., Universiti Teknologi Malaysia, 81310 Johor Bahru, Malaysia

Email: razak@utm.my

Abstract. The ever-growing demand of petroleum resources lead to depletion of easily accessible reservoirs. To meet the petroleum demands, exploration of challenging reservoirs must be taken up. This can be done by drilling deviated, horizontal and deeper well with high pressure and high temperature reservoirs. However, such drilling leads to harsh drilling circumstances such as torque and drag. These circumstances can be overcome by adding appropriate amount of nanomaterials to reduce drilling problems such as pipe sticking, thermal instability, torque and drag. This paper emphasis and investigates the usage of lubrication behaviour of drilling fluids by adding multi-walled carbon nanotubes (MWCNT) and graphene nano-platelets (GNP) to reduce coefficient of friction of water based drilling fluid. Experiments were conducted to study the characteristics of added nanoparticles on drilling fluid, where by analysing the rheological properties, filtrate loss fluid and lubricity. The results showed that the addition of MWCNT at 0.01 ppb and 0.02 ppb of GNP gave torque lubricity reduction between 38 to 59%. The comprehensive nanomaterial behaviour in water based drilling fluid favoured the enhancement of drilling fluid lubricity and marginally improved the drilling fluid rheological properties such as plastic viscosity, yield point and filtrate loss of drilling fluid.

1. Introduction

During extended and horizontal well drilling, one of the prime factors that limits the well length is the torque and drag between the drill string and wellbore. As a results, it decreases the drainage efficiency of oil reservoirs. Hook load increase during tripping due to high friction which leads to reduction in the equipment life and essentially decrease in the well length. There are several innovative drilling fluid systems that have been applied to combat these problems in completion and drilling engineering. One of such innovative drilling fluid system is obtained by adding appropriate amount of suitable nanomaterials in to the drilling fluid [1].

One of the major component of these torque and drag problems is the friction between the drill string and the borehole wall. Drilling fluid plays significant role in reducing the torque and drag problems. Micro and macro material-based drilling fluids have limited capability to reduce this torque and drag problems. Due to fine and very thin film forming capability of nanomaterial's, nano-based fluids can provide a significant reduction of the frictional resistance between the pipe and the borehole wall due to the formation of a continuous and thin lubricating film in the wall-pipe interface. Moreover, the tiny spherical nanoparticles may create an ultra-thin bed of ball bearing type surface between the pipe and the borehole wall and thus can allow easy sliding of the drill string along the nano-based ball-bearing

surface. This highlights the extraordinary role of nano-based smart fluid in reducing the torque and drags problems of horizontal, extended reach, multilateral and coiled tubing drilling [2].
The use of nanotechnology as an additive in drilling fluids has been investigated for several years resulting in mixed success. The necessity to either solve drilling problems or develop products that perform more efficiently has motivated fluids researchers to look towards nanotechnology and see what benefits it may bring in the area of rheology, fluid loss, reduction of coefficient of friction or improve lubricity of drilling fluids [3]. Recent works on addition of nano-additives in drilling fluid have proven increase in drilling fluid lubricity and improved rheological properties of water based drilling fluid and had been field trial successfully [4]. Taha *et al.* reported a reduction up to 80% in torque and an increase of 40 - 60% in rheological properties [4]. This graphene enhanced product provides superior lubricity and thermal stability to the water based drilling fluids will penetrate into microscopic pores of the tubular metal, crystalized in layers under high pressure, forming a protective film to improve lubricity, prevent bit balling, improve BHA's life span, improve ROP, and most important of all improving fluids' thermal stability.
The aims of this study are to compare and evaluate the coefficient of friction of the water based drilling fluid performance by adding multi-walled carbon nanotubes and graphene nano-platelets.

2. Methodology

Water based drilling fluid was prepared based on the composition shown in Table 1. The drilling fluid was weighed to achieve the density of about 13 lb/gal. MWCNT and graphene nano-platelets were added as additives at various concentration (0.01 - 0.04%). Tween-80 and distilled water were used to disperse the nano-materials to make sure that it will be uniform in the water based drilling fluid. All tests for drilling fluids were conducted according to standard test of API Recommended Practice 13B. The tests were conducted before hot rolled (BHR) and after hot rolled (AHR) at 250 ^{o}F for 16 hours.

Table 1. Composition of water based drilling fluid.

Material	Function
Distilled water	Based fluid
Potassium chloride (KCl)	Shale inhibitor
Caustic Soda (NaOH)	Alkalinity control agent
Flowzan	Viscosifier
Hydro-PAC UL	Filtrate control agents
Hydro-CAP XP (PHPA)	Retards dispersion of drill cuttings
Barite	Weighting agent

3. Results and discussion

Rheological properties play an important role in drilling fluid formulation. Addition of various concentration of nanomaterials show some improvements in the drilling fluid rheological properties as discussed below.

3.1. Plastic viscosity

The values of plastic viscosity without the addition of nano-materials, before and after hot rolled are 24 cP and 21 cP respectively. As shown in the Figures 1 and 2 for both nanomaterials, the trend shows marginal increment of plastic viscosity, but decrement upon adding various concentrations of GNP before hot rolled. MWCNT gives the higher values of plastic viscosity compared to GNP after hot rolled, mostly at 0.02% concentration. Nanoparticles consist of large surface areas per volume and it will increase the interaction of the nanoparticles with the matrix and surrounding water-based drilling fluid [5].

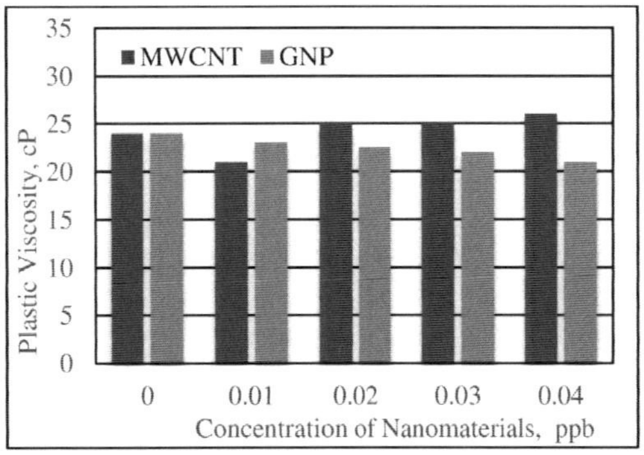

Figure 1. Plastic viscosity at different concentration of nanomaterials – before hot rolled.

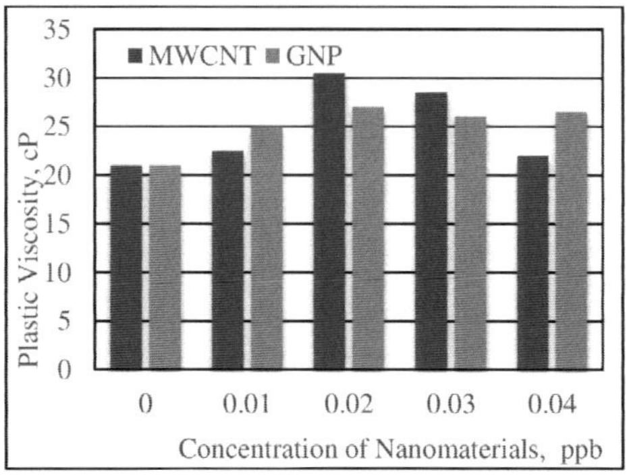

Figure 2. Plastic viscosity at different concentration of nanomaterials – after hot rolled.

3.2. Yield point

Figure 3 shows the effect of GNP and MWCNT concentration on yield point for before hot rolled. It can be seen that, yield point increases with increase in GNP concentration. Whereas, increasing the concentration of MWCNT the yield point shows unusual behaviour. After hot rolled, both the nanomaterials follow the same trend as shown in Figure 4. For MWCNT, yield point at 0.0 ppb is 32 $lb/100ft^2$ which increases to 36 at 0.01 and reaches maximum of 42 for 0.02 ppb. Upon further increase in concentration the yield point starts to decrease to 41 for 0.03 and 31 for 0.03.

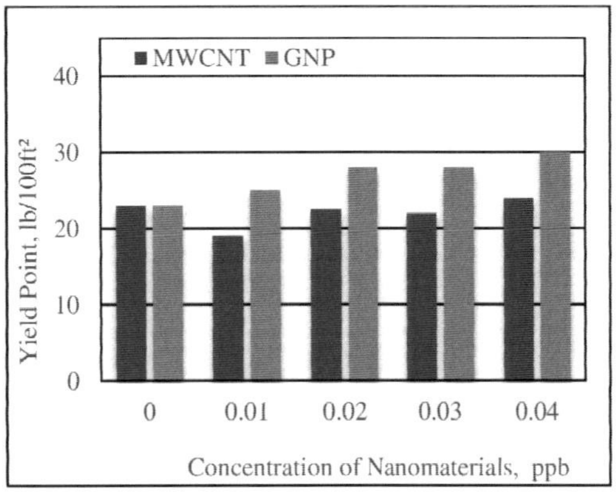

Figure 3. Yield point at different concentration of nanomaterials – before hot rolled.

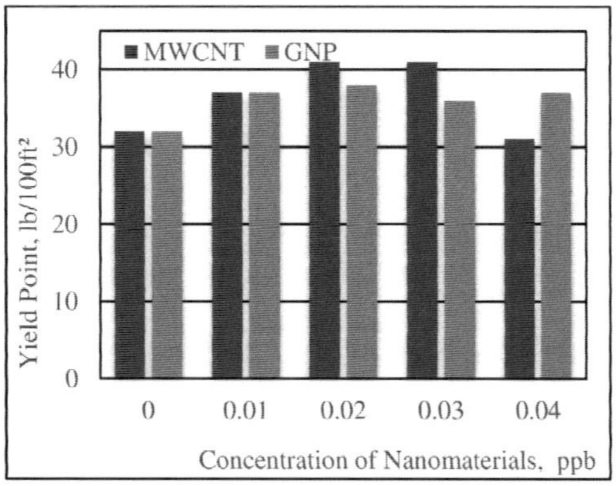

Figure 4. Yield point at different concentration of nanomaterials – after hot rolled.

3.3. Fluid loss or filtrate volume

Figures 5 to 8 showed that the fluid loss volume of MWCNT gives the lowest volume compared to GNP on both BHR and AHR, since MWCNTs have high surface area as it is hydrophobic (non-polar), which can prevent water from escaping the drilling fluid compared to GNP. In the shale formation it can prevent the swelling of shale. As per Ismail et al results showed that the filtrate loss of the water-based drilling fluid is reduced around 65% and filter cake thickness is reduced about 30% in the presence of 1 g of MWCNT [6]. These nanomaterials got attached to the drilling fluid material and formed thin and impermeable filter cake prevent fluid loss [7].

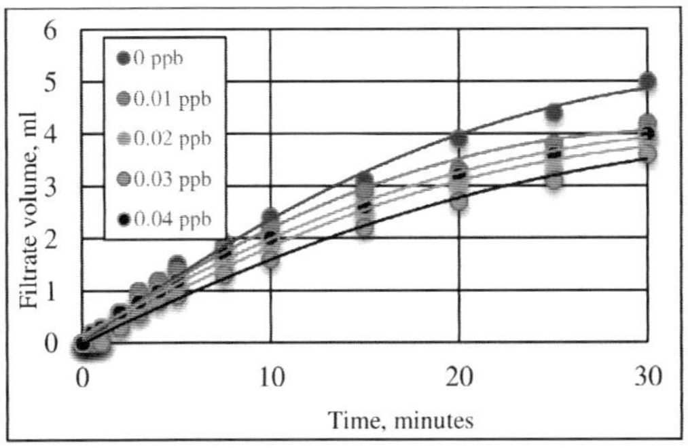

Figure 5. Fluid loss at different concentration of MWCNT – before hot rolled.

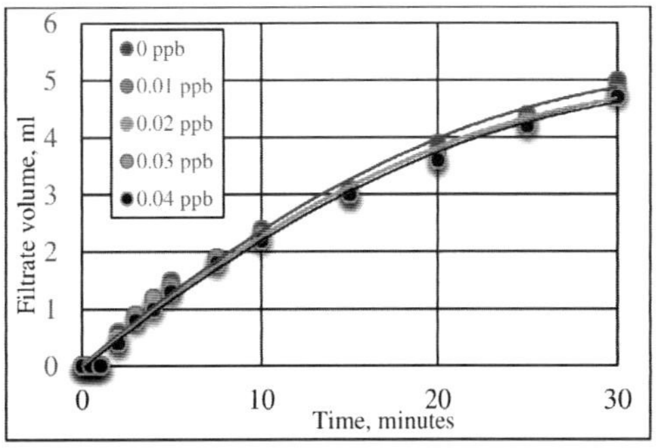

Figure 6. Fluid loss at different concentration of GNP – before hot rolled.

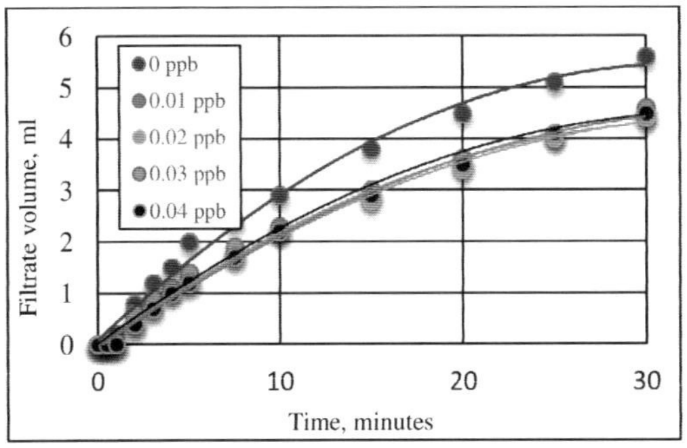

Figure 7. Fluid loss at different concentration of MWCNT – after hot rolled.

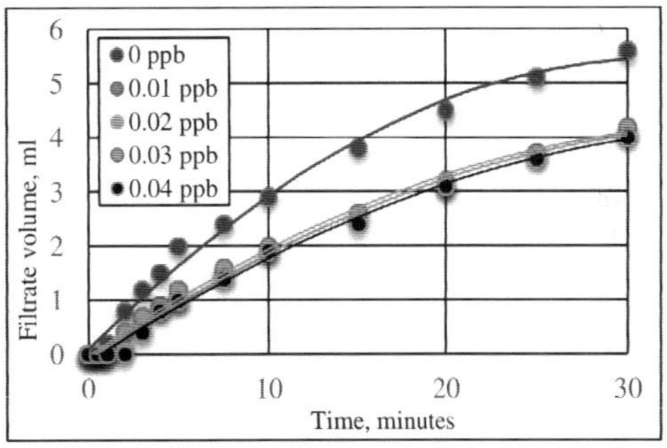

Figure 8. Fluid loss at different concentration of GNP – after hot rolled.

3.4. Lubricity

Coefficient of friction (CoF) describes the ratio of the force between two bodies and the force pressing them simultaneously. Figures 9 and 10 showed the CoF trends after addition of various concentration of nanomaterials into the drilling fluid, where MWCNT produces a reduction torque about 62% and GNP shows reduction torque of about 52% before hot rolled. GNP in water based drilling fluid at after hot rolled shoed slightly increased in CoF values that clarify the nanomaterial gives less effect to the lubricity of drilling fluid as concentration increased. Nonetheless, at concentration 0.01 ppb of MWCNT act as optimum concentration for enhance the torque reduction in WBM around 59% even at high temperature due to heat resistance capability.

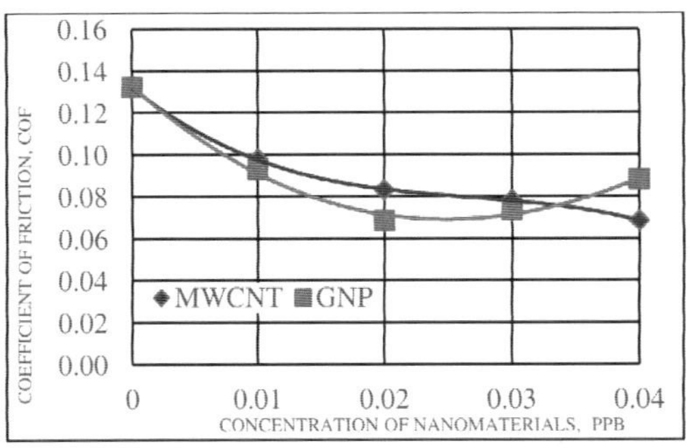

Figure 9. CoF at different concentration of nanomaterials – before hot rolled.

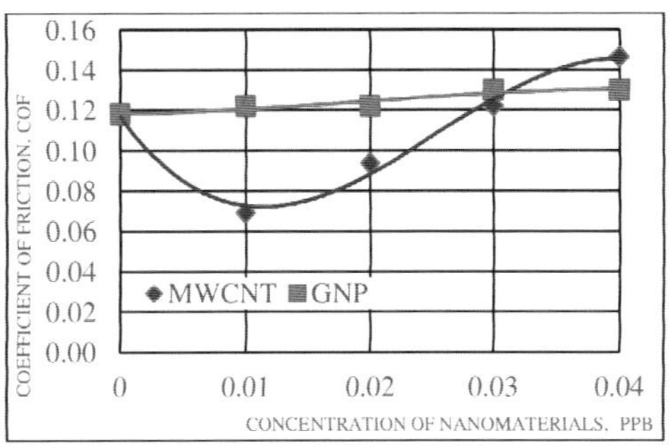

Figure 10. CoF at different concentration of nanomaterials – after hot rolled.

IOP Conf. Series: Materials Science and Engineering **380** (2018) 012021 doi:10.1088/1757-899X/380/1/012021

4. Conclusions

Based on the experimental study, several conclusions can be derived:

a) Addition of multi-walled carbon nanotubes and graphene nano-platelets in water based drilling fluid improved the rheological properties of the drilling fluid.
b) Addition of added multi-walled carbon nanotubes reduced fluid loss and promotes prevention of swelling in the shale formation.
c) Addition of a very small quantity of nanomaterials concentration resulted in drastic reduction of coefficient of friction from 38% to 59% before hot roll and the trend changed after hot roll as the nanomaterials get dispersed at high temperature.
d) Multi-walled carbon nanotubes provide better lubricity compared to graphene nano-platelets at elevated temperature.

5. Acknowledgments

The authors wish to thank Universiti Teknologi Malaysia for funding this project under Research University Grant vote Q.J130000.2546.17H77. The authors also appreciate Scomi Oiltools Sdn. Bhd. for providing the chemicals and additives to run the experiments.

6. References

[1] Geir H, Andrew W, Lingyun L, Husein M M and Muhammed F Z 2012 Innovative Nanoparticle Drilling Fluid and Its Benefits to Horizontal or Extended Reach Drilling, Alberta, Canada: *Society of Petroleum Engineers*, SPE 162686.

[2] Amanullah M, and Al-Tahini A M 2009 Nano-Technology - Its Significance in Smart Fluid Development for Oil and Gas Field Application, Al-Khobar, Saudi Arabia: *SPE Saudi Arabia Section Technical Symposium and Exhibition.*

[3] M-I SWACO 2009 M-I SWACO joins Rice University in Nanotech Research Program *Press Release.*

[4] Taha N M and Lee S 2015 Nano Graphene application improving drilling fluids performance *International Petroleum Technology Conference* IPTC-18539-MS

[5] Quintero L, Cardenas A E and Clark D E 2012 Nanofluids and Methods of Use for Drilling and Completion Fluids *US Patent 2012/0015852 A1.*

[6] Ismail A R, Seong T C, Buang N A and Sulaiman W R W 2014 Improve Performance of Water-based Drilling Fluids Using Nanoparticles *5th Sriwijaya International Seminar on Energy and Environmental Science and Technology*, Palembang, Indonesia.

[7] Sadeghalvaad M and Sabbaghi S 2015 The effect of the TiO_2/polyacrylamide nanocomposite on water-based drilling fluid properties *PowderTechnology* Vol 272 pp 113–119.

IOP Conf. Series: Materials Science and Engineering **380** (2018) 012022 doi:10.1088/1757-899X/380/1/012022

Nanosecond laser direct-part marking of data matrix symbols on titanium alloy substrates

C L Li

College of Mechanical and Automotive Engineering, Qilu University of Technology (Shandong Academy of Sciences), Jinan, Shandong, 250353, China

Abstract. In the application of laser marking, the biggest challenge is that machine-readable barcodes with superior quality were not marked consistently. To solve this problem, laser direct-part marking Data Matrix barcode experiments were carried out on titanium alloy substrates, using a Q-switched light-pumped Nd:YAG laser. The microstructure of the symbols was analyzed using an environmental scanning electron microscope (ESEM). The internal micro-stresses of the marked areas were analyzed using X-ray diffractometer (XRD). The influence of the pulse frequency on the symbol contrast is analyzed. Results showed the interaction between the laser and the titanium alloy can be found. This can further explain the physical mechanism of laser direct part marking Data Matrix symbols on titanium alloy substrates.

1. Introduction

In a harsh environment, the two-dimensional (2D) barcode symbol attached to an item is not difficult to be damaged. This is resolved with an approach known as laser direct part marking, which refers to the use of laser thermal effect to produce a permanent mark on the surface of an item. When a laser beam contacts with an item, its light energy is converted into heat energy, which creates a mark on the surface of the item. The mark visibility can be achieved by surface material removal or color change. The difference between the two methods to produce a permanent mark is the average power of the laser used. Surface material removal in laser direct part marking includes laser etching and laser engraving which requires a higher power than laser etching. Whereas color change refers to laser coloring which requires a lower power than laser etching. The intense laser beam may change the color of an item. Meanwhile the item may be damaged owing to the heat effect of the laser which must be avoided. Various materials may react differently to each type of laser marking technique. Owning to the complexity of the interaction between laser and materials, the industrial application of laser marking Data Matrix symbols was affected greatly. A direct effect of the interaction between laser and material is the material damage irradiated by laser. The damage is linked to the laser parameters and the characteristic of the substrate material. So the research on the physical mechanism of laser marking can be transformed into interaction process analysis between laser and materials.

The physical process during laser-material interaction involves many aspects such as heating, absorption, melting, evaporation, reflection, propagation, recoil pressure, piston effect, plasma formation, etc. When the laser energy was absorbed, the subsequent effect will arise. The classical laser physical mechanism was utilized during the investigation of the physical mechanism of laser marking [1-3]. Laser melting [2], ablation [3], plasma and etc. Various laser ablation models were proposed in recent years. Peligrad put forward a melt depth prediction model for quality control of laser surface glazing of inhomogeneous materials [4]. Also Peligrad proposed a muti-input and multi-

IOP Conf. Series: Materials Science and Engineering **380** (2018) 012022 doi:10.1088/1757-899X/380/1/012022

output dynamic model for laser marking of ceramic materials [5]. But research on the physical mechanism of Nd:YAG laser marking 2D barcodes onto titanium alloy has not been reported.

2. Experimental details

2.1. Experimental materials and setup

The substrate materials used throughout this experiment are three TC4 titanium alloy sheets. One is 200 mm×200 mm×2 mm. The other two are 50 mm×50 mm×2 mm. The experimental setup is a solid state laser, Nd:YAG laser, whose applied processing characteristics are shown in Table 1.

Table 1. Processing characteristics of the laser marking equipment.

Processing characteristics	Value
Laser wavelength	1064nm
MAX Q-switch laser power	70W
Pulse duration	80~260ns
Beam quality	M2<10
Marking line width	<150μm
Marking depth	<2.0mm

There are many kinds of 2D barcodes, from which the Data Matrix symbol is selected in the experiment. It encodes the data string 1A2B3C4D5E. Its dimensions are 16×16 modules. The size of the obtained Data Matrix symbol onto titanium alloy substrates is 10×10mm.

Some marked Data Matrix symbols were analyzed using a D8-ADVANCE X-ray diffractometer (XRD), which is copper (Cu) target X-ray tube and θ/θ goniometer. The sample remains level during measurements. The angle repeatability of the goniometer is 0.0001° and its degree of accuracy is ±0.001°. The minimum angle is 0.2 °. The XRD is produced by Bruker AXS Company in Germany. Moreover the mark was measured utilizing a Quanta 200 environmental scanning electron microscope (ESEM) whose resolution is 3.5nm.

2.2. Experimental procedure

A Data Matrix symbol was marked in the center of one TC4 titanium alloy when the current intensity is 28A, the pulse frequency is 3 kHz, the filling space is 0.01 mm, the scanning speed is 16.7mm/s. Then a Data Matrix symbol was marked in the center of the other titanium alloy sheet when the current intensity is set to 16A, while other experimental parameters being unchanged. The ESEM photographs (25×) of the Data Matrix symbols for different current intensity are shown in Figure 1 (a) and (b).

In order to compare the effect which laser direct part marking Data Matrix symbols induces on the titanium alloy surface, three samples were prepared. They are unmarked sample, sample marked with 16A and sample marked with 28A. XRD analysis has been used to detect micro-stresses in the marked areas of substrate materials. XRD analysis results are presented in Figure 2.

Eighteen Data Matrix symbols were marked when the pulse frequency is 1, 2, 3, 4, 5, 6, 7, 8, 9,10, 12, 15, 17, 19, 22, 28, 34, 40 kHz respectively. Fourteen symbols are marked when the pulse frequency is within its operating range and four symbols are marked when the pulse frequency is beyond its operating range. The images of representative Data Matrix symbols are given in Figure 3 (a), (b), (c) and (d).

IOP Conf. Series: Materials Science and Engineering **380** (2018) 012022 doi:10.1088/1757-899X/380/1/012022

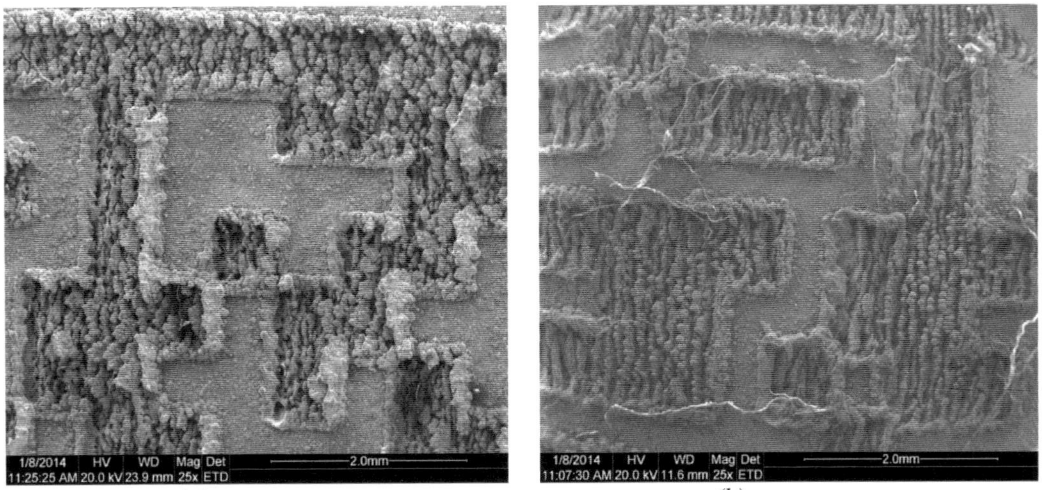

(a) (b)

Figure 1. ESEM photographs (25×) of the Data Matrix symbols for different current intensity: (a) 16A. (b) 28A.

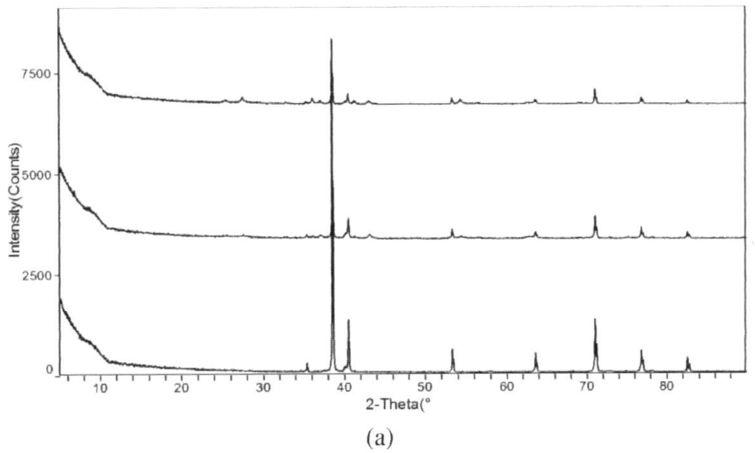

(a)

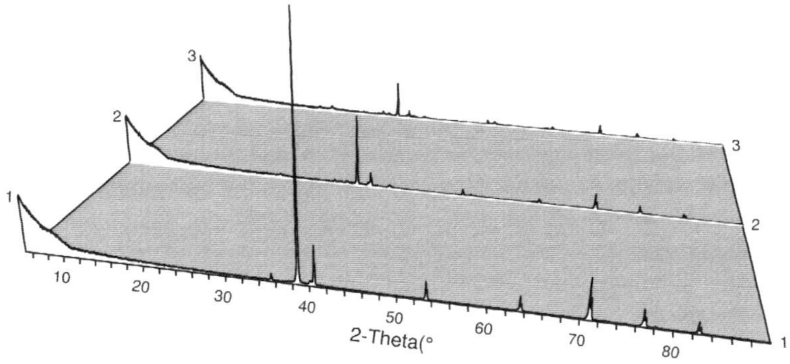

(b)

Figure 2. X-ray diffraction graphs for θ in the range of 0-90° (a) a two-dimensional representation (b) a tridimensional representation: (1) unmarked sample. (2) sample marked with 16A. (3) sample marked with 28A.

(a) (b) (c) (d)

Figure 3. Image of laser marked Data Matrix symbols for different pulse frequency: (a) 12 kHz. (b) 15 kHz. (c) 22 kHz. (d) 40 kHz.

3. Results and discussions

Figure 1(a) and (b) shows the marked area is more rough than unmarked area and some dust residues were not been removed completely. Leone [6] studied the relationship between the symbol contrast and roughness. The symbol contrast increases with roughness when it is less than 4μm, beyond this limit, the symbol contrast is independent of roughness. In Figure 1(a) and (b), the roughness of the marked areas are all more than 4μm, so the symbol contrasts are not much different.

As seen in Figure 2, there is no appearance of new diffraction peaks or increase in the width of the diffraction peaks in the marked areas. So there are not any modifications in the nature of the phases and its structure. Mainly, the microstructure is not affected. There is no displacement of diffraction peaks initially recorded on the unmarked material. Therefore, one can say that there are no internal stresses as a consequence of laser marking Data Matrix symbols on substrate materials. In all, it is feasible for titanium alloy substrates to be marked Data Matrix symbols by nanosecond laser.

The symbol contrast of the Figure 3(a) (b) (c) and (d) is 0.2807, 0.4818, 0.7107 and 0.6860 respectively. In fact, the two marks (a) and (b) are obtained by laser etching which involves material removal, leaving an engraved mark. Whereas the two marks (c) and (d) are produced by laser discoloration which uses a lower energy density than laser etching, leaving a flush and smooth mark. Laser discoloration involves the superficial oxidation of certain elements, but it didn't involve material removal, Variations in the color change may be achieved by varying the laser parameters, and a variety of cosmetic effects can be obtained.

4. Conclusions

A nanosecond laser was utilized to mark several Data Matrix symbols on titanium alloy substrates in the experiment. Owing to the fact that laser marking involves material removal, the marked area is more rough than unmarked area. Laser discoloration can also produce a permanent mark, except for laser etching which requires a higher power than laser discoloration. In order to compare the effect which laser marking induces on the substrate material, three samples were prepared. They are unmarked sample, sample marked with 16A and sample marked with 28A. XRD analysis results show the microstructures of the marked substrate materials are not damaged and there is no internal stresses as a consequence of laser marking. So it is safe for titanium alloy substrates to be marked by nanosecond laser.

5. Acknowledgments

This work was financially supported by National Natural Science Foundation (Funding No: 51375275).

6. References

[1] Sanner N, Huot N and Audouard E 2007 Direct ultrafast laser micro-structuring of materials using programmable beam shaping *Opt. Laser Eng.* **45** pp 737-41

[2] Valette S, Audouard E and Harzic R L 2005 Heat affected zone in aluminum single crystals submitted to femtosecond laser irradiations *Appl. Surf. Sci.* **239** pp 381-6
[3] Yilbas B S, Shuja S Z and Khan S M A 2011 Laser repetitive pulse heating of tool surface *Opt. Laser Technol.* **43** pp 754-61
[4] Peligrad A A, Zhou E and Morton D 2001 A melt depth prediction model for quality control of laser surface glazing of inhomogeneous materials *Opt. Laser Technol.* **33** pp 7-13
[5] Peligrad A A, Zhou E and Morton D 2001 A multi-input and multi-output dynamic model for laser marking of ceramic materials *J. Laser Appl.* **14** pp 31-8
[6] Leone C, Genna S and Caprino G 2010 AISI 304 stainless steel marking by a Q-switched diode pumped Nd:YAG laser *J. Mater. Process Tech.* **210** pp 1297-303

IOP Conf. Series: Materials Science and Engineering **380** (2018) 012023 doi:10.1088/1757-899X/380/1/012023

Research on the quality of laser marked data matrix symbols

C L Li

College of Mechanical and Automotive Engineering, Qilu University of Technology (Shandong Academy of Sciences), Jinan, Shandong, 250353, China

Abstract. To evaluate the quality of the laser direct part marked Data Matrix symbols on titanium alloy substrates, the quality assessment methods at home and abroad were compared. A new quality assessment method of combining the effect of the laser on substrate materials and symbol grade of laser marked Data Matrix was put forward. Depending on previous research works, orthogonal experiment results were analyzed again and a modified nonlinear mathematics model was established. Analysis results indicate that this modified model can explain 90.6% of symbol contrast change and it is statistically significant. So it is better than previous linear regression model and can be used to estimate the quality of laser marked Data Matrix symbols on titanium alloy substrates. The nonlinear mathematics model can also explain the laser parameters influence on the symbol contrast.

1. Introduction

There is a growing need in the industry to track and trace product during production or during its lifetime. As the product identification is required to last its entire lifetime, the marking must be permanent. Laser marking can generate permanent marks on a wide variety of materials. In the entire supply chain, two-dimensional barcode which is the carrier of information has played an important part all along. So more and more people start to pay close attention to the quality of the barcodes. A Data Matrix has many advantages over other barcodes. For example, it can encode a lot of information in the smallest space and its error correction provides safeguards for poorly printed or damaged symbols. So it is the most popular two-dimensional barcode. In industry, many associations select Data Matrix in product tracking and the quality of the laser marked Data Matrix symbol has become the focus of research.

At present, there is no unified quality evaluation method. Some scholars such as Sheu [1] and Hayakawa [2] use barcode identification equipment to evaluate the quality of laser marked Data Matrix, they utilize readability characteristic as a quality indicator. The functions and characteristics of barcode identification equipment are different and the manufacturers try to improve their decoding algorithm to maximize the performance of their product. Thus the barcode identification equipment may decode some poorly printed Data Matrix. So this approach is not accurate enough. Some scholars such as Qi et al. [3] and Ba [4] and Leone et al. [5] use mark contrasts, mark depth, mark width, and surface roughness in the marked area as the quality indicator. The mark contrast is one of five performance measures (i.e. decode, symbol contrast, print growth, axial nonuniformity and unused error correction) which are specified in ISO/IEC 16022 standard [6]. Jangsombatsiri et al. [7] use the five performance measures to assess the quality of laser marked Data Matrix on carbon substrates. But ISO/IEC 16022 standard is set for paper-printed two-dimensional Data Matrix. Whether this standard is applicable to laser marked Data Matrix is not known yet. Qiu [8] use symbol grades measured by barcode verifier to assess the quality of laser marked Data Matrix on aluminum substrates. Barcode

IOP Conf. Series: Materials Science and Engineering **380** (2018) 012023 doi:10.1088/1757-899X/380/1/012023

verifier adopted standardized reference decoding algorithm, and the opticator was calibrated, so it can make a unified objective evaluation about the quality of the barcode. But the process of laser direct part marking is thermal effect. Heat affected zone can't be completely eliminated. In addition, the formation of micro cracks and recast layer will affect the physical properties of the base material. There might provide for a situation when a laser marked Data Matrix symbol grade is A, but the substrate material deformation occurs. So symbol grade measured by barcode verifier cannot stand for the quality of laser marked Data Matrix fully. Therefore the quality of laser marked Data Matrix symbol can be judged by the effect of laser on base materials and symbol grade.

The author has studied the effect of laser on base material in previous research, which showed that deformation and crack did not exist in substrate materials after laser marking [9]. One can say that the effect of laser on base materials is negligible in this research. According to ISO/IEC 16022 standard[6] and previous research [10], the symbol grade for a Data Matrix symbol can be determined by symbol contrast. Whereas the symbol contrast is closely related to the laser parameters. So the aim of this study is to investigate the relationship between symbol contrast and laser parameters.

2. Experimental details

2.1. Experimental setup and materials

A neodymium-doped, yittrium-aluminium-garnet (Nd :YAG) laser was used in the study, whose pumping source is krypton lamp. The laser operated at 80W MAX average power, wavelength of 1064nm, operating frequency of 0-20kHz, pulse width of 80~260ns, and laser spot diameter of 0.010~0.150mm [10]. The marking depth is less than 2.0mm [11]. The substrate materials used in this study are TC4 titanium alloy sheets.

2.2. Reference Data Matrix Symbol

The reference Data Matrix symbol encodes the data string 1A2B3C4D5E and its dimensions are 16×16 modules [9-12]. An example of a reference Data Matrix symbol is given in Figure 1 (a). The size of a laser maked Data Matrix symbol is is 10×10mm and its image is given in Figure 1 (b).

(a) (b)

Figure 1. a) Example of a reference Data Matrix symbol, b) Image of a laser marked Data Matrix symbol.

2.3. Experimental procedure

OFAT experimental results showed that the current intensity, effective vector step, Q-switch frequency, and laser line spacing had a statistically significant effect on the contrast of laser marked Data Matrix symbols on titanium alloys[12]. Based on this conclusion a set of orthogonal experiments were carried out, the experimental results were analyzed with the method of analysis of variance(ANOVA) and signal to noise ratio analysis[11], and a multiple linear regression model was constructed[10]. But the interpretative capability of the model is just 79.5% which is not very good. Based on orthogonal experimental results, interactions plot matrix for symbol contrast is shown in Figure 2.

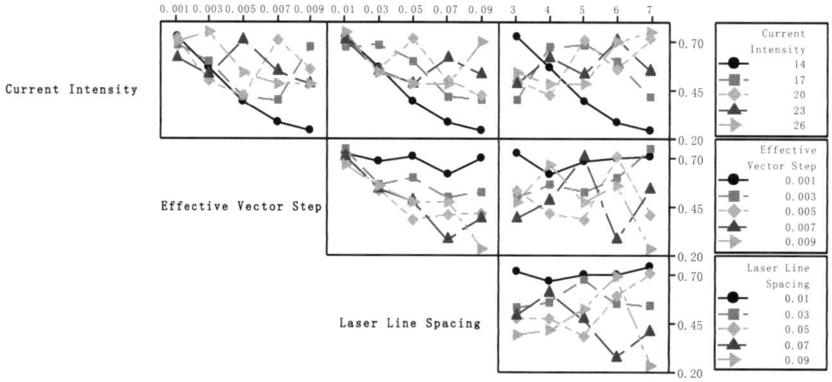

Figure 2. Interactions plot matrix of laser parameters.

Figure 2 indicates the interactions plot matrix of laser parameters. There are obvious interactions between the laser parameters. From this figure, one can say that the relationship between the symbol contrast and laser parameters are nonlinear. So nonlinear mathematical model is more suitable in interpreting the experimental data.

3. Development a regression model

For the convenience of analysis in SPSS software, nonlinear problems should be converted to linear problems. The regression analysis and ANOVA results are shown in Table 1 and the following nonlinear mathematics model was obtained based on Table 1.

$$y = 0.659 - 323.251 \cdot b \cdot c - \frac{30.704}{a^2} + \frac{1.02\times10^{-7}}{b^2} + \frac{1.347\times10^{-5}}{c^2} \quad (1)$$

Where, y is the symbol contrast, a is the current intensity, b is the effective vector step and c is the laser line spacing.

Table 1. Regression Analysis and ANOVA Results.

Parameters	Coefficient	t	p
(Constant)	0.659	22.895	0.000
b • c	-323.251	-5.491	0.000
a^{-2}	-30.704	-4.433	0.000
b^{-2}	1.020E-7	3.630	0.002
c^{-2}	1.347E-5	4.794	0.000
ANOVA	SS	F	p
Regression	0.447	58.992	0.000
Model Summary	R	R Square	Adjusted R Square
	0.960	0.922	0.906

The coefficient in Table 1 indicates the coefficient values of variables in the nonlinear mathematics model. The subsequent p indicates whether there is a significant correlation between each coefficient toward a parameter and symbol contrast. If the p value is lower than 0.05, then the parameter is significant. The p value in ANOVA results in Table 1 shows the developed mathematics model is statistically significant. The model summary in Table 1 indicates the model fitting. It represents interpretative capability of the model. The bigger the adjusted R square, the greater the interpretative capability of the model. The adjusted R square in Table 1 is 0.906 which illustrate the model can explain 90.6% of symbol contrast change. The predicated symbol contrast value can be obtained when the 25 orthogonal experimental conditions were inserted into the nonlinear mathematics model. Results were shown in Table 2. Comparing results of experimental measured value and predicted value for symbol contrast are given in Figure 3.

Table 2. The predicated symbol contrast value.

No.	Predicted symbol contrast value
1	0.7361
2	0.4997
3	0.4312
4	0.3490
5	0.2436
6	0.6602
7	0.5212
8	0.4466
9	0.3530
10	0.6598
11	0.6737
12	0.5286
13	0.4427
14	0.6966
15	0.5114
16	0.6833
17	0.5268
18	0.7238
19	0.5503
20	0.4623
21	0.6884
22	0.7501
23	0.5843
24	0.5081
25	0.4141

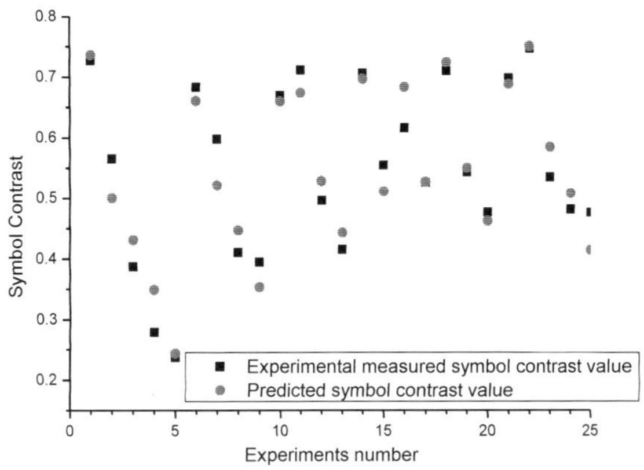

Figure 3. Comparing results of experimental measured value and predicted value for symbol contrast.

From Table 2 and Figure 3, one can say that the experimental measured value and predicted value for symbol contrast is very close. So the modified nonlinear mathematics model can be used to estimate the quality of laser marked Data Matrix symbol.

4. Conclusions

To evaluate the quality of the laser marked Data Matrix symbols, the research status at home and abroad was reviewed. A further quality assessment methods of combining the effect of the laser on base materials and symbol grade was put forward. Previous works have demonstrated that the effect of laser on base materials is negligible and the symbol grade for a Data Matrix symbol can be determined by symbol contrast. Whereas the symbol contrast is closely related to the laser parameters. So previous orthogonal experiment results were analyzed again and a modified nonlinear mathematics model was established. Analysis results indicate that this modified model can explain 90.6% of symbol contrast change and it is statistically significant. So it is better than previous linear regression model and can be used to assess the quality of laser marked Data Matrix symbol. This can also explain the influence of laser parameters and substrate materials on the symbol contrast.

5. Acknowledgments

This work was financially supported by National Natural Science Foundation (Funding No: 51375275).

6. References

[1] Daniel Sheu D 2000 A laser marking system for flexible circuit identification *J. Manuf. Syst.* **19** 202-12

[2] Hayakawa H 2000 A laser method for marking bar codes on glass substrates *Proc. SPIE* 4088 363-6

[3] Qi J, Wang K L and Zhu Y M 2003 A study on the laser marking process of stainless steel *J. Mater. Process Tech.* **139** 273-6

[4] Ba F 1998 Technical Study on Laser Marking of Slide Calipers *Proc. SPIE* **3550** 336-41

[5] Leone C, Genna S and Caprino G 2010 AISI 304 stainless steel marking by a Q-switched diode pumped Nd:YAG laser *J. Mater. Process. Tech.* **210** 1297-303

[6] ISO, 2004, ISO/IEC 16022: 2000 Information technology-Automatic identification and data capture techniques-Bar code print quality test specification-Two dimensional symbols, 1st ed., International Organization for Standardization, Geneva, Switzerland.

[7] Jangsombatsiri W and Porter J D 2007 Laser direct-part marking of data matrix symbols on carbon steel substrates *J. Manuf. Sci. Eng.* **129** 583-91

[8] Qiu H D, Lu C H and Xiao R J 2011 Laser parameters influence in direct-part marking barcodes on aluminium alloy *Chin. J. Lasers* **38** 95-9

[9] Li C L 2014 Study of effect of laser marking data matrix symbols on titanium alloy sheets using Nd:YAG laser *Adv. Mater. Res.* **941-944** 2165-68

[10] Li C L 2014 Contrast prediction for laser direct part marked data matrix symbols on titanium alloy substrates *Adv. Mater. Res.* **941-944** 2161-64

[11] Li C L 2014 Impact of laser parameters on the contrasts of laser direct marked data matrix symbols on titanium alloy substrates *Adv. Mater. Res.* **915-916** 1018-22

[12] Li C L 2014 OFAT experimental study of laser direct-part marking of data matrix symbols on titanium alloys *Adv. Mater. Res.* **915-916** 1027-31

AUTHOR INDEX